Communications
in Computer and Information Science 282

Dieter Uckelmann Bernd Scholz-Reiter
Ingrid Rügge Bonghee Hong
Antonio Rizzi (Eds.)

The Impact of Virtual, Remote, and Real Logistics Labs

First International Conference, ImViReLL 2012
Bremen, Germany, February 28 – March 1, 2012
Proceedings

Springer

Volume Editors

Dieter Uckelmann
University of Bremen, LogDynamics Lab
28359 Bremen, Germany
E-mail: dieter@web-of-things.com

Bernd Scholz-Reiter
BIBA GmbH
28359 Bremen, Germany
E-mail: bsr@biba.uni-bremen.de

Ingrid Rügge
University of Bremen
International Graduate School for Dynamics in Logistics
28359 Bremen, Germany
E-mail: rue@biba.uni-bremen.de

Bonghee Hong
Pusan National University
Institute of Logistics Information Technology
Pusan 609-735, Korea
E-mail: bhhong@pusan.ac.kr

Antonio Rizzi
University of Parma, RFID Lab
43100 Parma, Italy
E-mail: antonio.rizzi@unipr.it

ISSN 1865-0929 e-ISSN 1865-0937
ISBN 978-3-642-28815-9 e-ISBN 978-3-642-28816-6
DOI 10.1007/978-3-642-28816-6
Springer Heidelberg Dordrecht London New York

Library of Congress Control Number: Applied for

CR Subject Classification (1998): J.1, H.4, C.3

Typesetting: Camera-ready by author, data conversion by Scientific Publishing Services, Chennai, India

Printed on acid-free paper

Springer is part of Springer Science+Business Media (www.springer.com)

Preface

This volume contains the proceedings of the first international conference on "The Impact of Virtual, Remote and Real Labs in Logistics" 2012 (ImViReLL 2012). The conference took place in combination with the "International Conference of Dynamics in Logistics" 2012 (LDIC 2012) in Bremen, Germany.

The importance of logistics labs in supporting the transition from fundamental to applied research is undisputed. Lately, the Internet (of Things) has changed the scope of these labs. Firstly, the Internet of Things is addressed as a research topic in these labs. Secondly, the Internet supports virtualization of testing infrastructures and remote access to logistics labs for increased collaboration between researchers across different research disciplines and locations. The goal of the conference was to focus on lab-based research and education, evaluate their impact in research and education and investigate specific demands, opportunities and challenges.

The idea to have a conference focusing on the role of logistics labs and their relevance for research and education has largely been influenced by corresponding research on labs in other disciplines, such as chemistry, electronics, engineering and automation. Collaborations, such as the Virtual and Remote Labs Community (www.vrlcom.com), lab2go (www.lab2go.net), Global Online Laboratory Consortium (online-lab.org) as well as conferences (e.g., Remote Engineering and Virtual Instrumentation – REV) and journals, such as the *International Journal of Online Engineering – iJOE*, have provided different platforms to exchange ideas and research results about lab-based research. However, logistics labs have spread during the last years without an established information exchange community.

The ImViReLL conference tries to fill this gap. On the one hand, it provides a platform for information exchange between lab researchers in logistics. It thus supports and complements the work of the Global RF Lab Alliance (www.grlfa.org), RFID in Europe and the *International Journal of RF Technologies*, which are focused on radio-frequency-based research and applications in logistics. On the other hand, ImViReLL invites researchers from other disciplines to support interdisciplinary research and information exchange. This approach is not without problems, however. Some reviews, for example, have been declined by scientific committee members from other disciplines if papers were too domain-specific to logistics. It will take further efforts to differentiate domain-specific from domain-independent topics.

ImViReLL 2012 provided a venue for researchers from academia and industry interested in future-oriented logistics labs as a nucleus for innovation. The conference covered research in logistics from a wide range of fields, for instance, engineering, computer science, distributed education and collaborative research.

The conference addressed numerous specific areas of interest, such as lab-based technology and feasibility studies, pilots and demonstrators, lab-centric specifics of logistic labs, virtual and remote research environments and communities, the role of RFID, sensors, actuators, robots, intelligent material handling, and (de-) centralized data processing in logistic labs, architecture developments for multiple (networked) demonstrators, social networking technology in research and educational implications. This diversity of topics is reflected in the conference papers.

We would like to thank all authors, Scientific Committee members and reviewers for their work. Additionally, we would like to thank RFID in Europe, Concord, LogDynamics, the Global RF Lab Alliance and *RFID Journal* for their helpful support.

March 2012

Dieter Uckelmann
Bernd Scholz-Reiter
Ingrid Rügge
Bonghee Hong
Antonio Rizzi

Organization

Conference Chairs

Dieter Uckelmann	LogDynamics Lab, University of Bremen, Germany
Bernd Scholz-Reiter	University of Bremen, Germany
Ingrid Rügge	University of Bremen, Germany
Bonghee Hong	Institute of Logistics Information Technology (LIT), Pusan National University, Korea
Antonio Rizzi	RFid Lab, University of Parma, Italy

Scientific Committee and Reviewers

Gisele Bennett	Electro-Optical Systems Laboratory, Georgia Tech Research Institute, USA
Shing-Chi Cheung	RFID Center, Hong Kong University of Science and Technology, China
Volker Coors	HFT Stuttgart, Germany
Farideh Ganji	BIBA GmbH, Germany
Javier García-Zubia	Deusto Institute of Technology, Spain
Christian Gorldt	BIBA GmbH, Germany
Willibald Günthner	TU Munich, Germany
Bill C. Hardgrave	Auburn University, Bremen
Hamid Reza Karimi	University of Agder, Norway
Dietmar Kennepohl	Athabasca University, Canada
Michael Lawo	University of Bremen, Germany
Yu Liu, RFID Lab	Chinese Academy of Science (CASIA), Beijing, China
José Machado	University of Minho, Portugal
Thomas Makuschewitz	BIBA GmbH, Germany
Florian Michahelles	Auto-ID Lab, ETH Zurich, Switzerland
Kaj Nummila	VTT Technical Research Center of Finland
Justin Patton	University of Arkansas RFID Research Center, USA
Katerina Pramatari	ELTRUN Laboratory (AUEB), Athens University of Economics and Business, Greece

Michael Schenk	Fraunhofer-Institut für Fabrikbetrieb und -automatisierung IFF, Magdeburg, Germany
Samuel Bloch da Silva	Flextronics Institute of Technology – RFID CoE, Brazil
Klaus-Dieter Thoben	University of Bremen, Germany
Keith Ulrich	DHL Solutions & Innovations, Germany
Markus Witte	Lufthansa Cargo AG, Germany

Local Organization

Aleksandra Himstedt	University of Bremen, Germany
Irena Weinhold	University of Bremen, Germany

Table of Contents

Fundamentals and Historic Background of Lab-Based Research in Logistics

Infrastructure and Design of Virtual, Remote and Real Labs

Educational Implications of Virtual, Remote and Real Labs

Test-Beds and Demonstrators

Lab-Based Process Improvements in Logistics

Lab-Supported Product Developments

The Role of Logistics Labs in Research and Higher Education

Dieter Uckelmann

LogDynamics Lab, University of Bremen,
Hochschulring 20, 28359 Bremen, Germany
uck@biba.uni-bremen.de

Abstract. Lab-based logistic research serves multiple goals. It bridges the gap between fundamental and applied research, it is used as a demonstration infrastructure for new developments, thus supporting business innovation, and it supplies training and participation opportunities for students as well as for practitioners. No wonder that numerous university and industrial logistics research labs exist. However, the role of logistics labs for research and education has not yet been researched in detail and the transition from in-house logistics to supply networks research has not yet been transferred into logistics labs. Additionally, new developments, such as the Internet of Things and the Web 2.0, are changing the focus of logistics labs and increase the importance of interdisciplinary research with computer science faculties.

Keywords: Logistics Research, Internet of Things, Logistic Laboratories, Information Logistics.

1 Introduction

This paper takes a closer look at the history and future of logistics labs. First, the historic developments of logistics research and education will be discussed with a focus on logistics education at university level in Germany. It will be shown that while logistics labs exist since the 1950's, there has not been a deeper investigation on lab-based logistics research ever since. Instead, other research disciplines have researched topics such as virtual research environments and remote engineering, which are, however, just as relevant to logistics. In the following, a view on the relevant components in logistics labs is given and the increasing role of information technology is highlighted. The importance of computer sciences in logistics is not yet sufficiently represented in research and education. Instead, logistics research is mainly divided between business economic and engineering disciplines. With the increasing multidisciplinary of logistics research and education, it will be more important to allow non-engineering disciplines access to lab-based research infrastructures to verify and demonstrate research approaches in realistic environments, as logistics research by default is considered a discipline of applied science. Additionally, the gap between universities that run a logistics lab at tight budgets and those that have no access to corresponding infrastructures, due to cost and/or space reasons, needs to be bridged.

Dieter Uckelmann et al. (Eds.): ImViReLL 2012, CCIS 282, pp. 1–12, 2012.

Remote access may achieve a better utilisation and financing of lab infrastructures, while providing access to a broader group of logistics researchers. Logistics labs that provide remote access will also have the opportunity to connect in order to do research on networked logistics and information infrastructures, which may be considered a prerequisite for a lab-abstraction of supply networks in the real world.

2 Historic Developments and Current State of Logistics Research, Logistics Education, and Logistics Labs

While the origins of the term logistics may be tracked back to the ancient Greeks [1], its first mentioning as an 'art of war' relates to the Byzantine emperor Leo (Leontos) VI (886-911, surnamed "the wise" or "the philosopher") [2]. The civil research field of logistics only dates back to the last century, when the term was taken over from military strategies by universities after the Second World War [2]. However, there were already some prior publications from marketing disciplines that mentioned the importance of physical distribution [3]. Ihde [2] identifies a publication by Morgenstern [4] as the initiation point for logistics as a scientific discipline. Ballou [3] considers a publication by Smykay, Bowersox and Mossmann [5] on "Physical Distribution Management" as the first relevant textbook to the field. The distribution aspect was complemented by the supply side and a focus on the business aspects in Heskett, Ivie and Glaskowsky's [6] "Business Logistics: Management of Physical Supply and Distribution". The term logistics has been complemented by the term supply chain management to reflect the interdependency between different companies and stakeholders. Thus, the logistics research discipline has gone through several transition periods and will continue to do so. As an example, the National Council of Physical Distribution Management (NCPDM) founded in 1963, changed its name to Council of Logistics Management in 1985 and, consequently, to Council of Supply Chain Management Professionals in 2004 [7].

Germany is one of the key-players in logistics and has been ranked number one in the Logistics Performance Index (LPI) 2010 [8]. Today, logistics is considered to be the third biggest economic sector with a volume of about 211 billion Euro in 2011 in Germany [9]. However, until today logistics research is still on its way of being recognized as an independent research discipline, for example, through establishing corresponding logistics faculties. Even dedicated logistics curricula are scarce. Baumgarten and Hildebrand [10] found that students studying logistics in Germany did choose between dedicated logistics curricula (11% of alumni in 2007), logistics as a major field of study (30% of alumni in 2007) and logistics as part of another study focus (59% of alumni in 2007), which may be largely effected by the individual offerings at the universities and thus does not represent a free choice. In a review of university websites in Germany, on average, 27% of the achievable credit points in logistics curricula could be directly associated to logistics. The renowned CHE-ranking in Germany lists 37 different disciplines – but does not mention logistics [11]. Craighead, Hanna, Gibson, and Meredith [12] emphasize the acceptance problem of logistics in academic disciplines as follows:

"While it is clear many practitioners are focused on logistics and supply chain issues, the academic segment of these disciplines, while making significant progress, sometimes struggles for broad acceptance."

The first mentioning of a logistics lab dates back to Haythorn's [13] "Simulation in Rand's logistics systems laboratory". This Logistics Systems Laboratory (LSL) was initiated under US Air Force sponsorship in 1956 [14]. Harman [14] described the lab as follows:

"The system consisted of men and machine resources together with policy rules on the use of such resources in simulated stress situations such as war."

This shows quite clearly that laboratory research has played an important role in logistics for more than 50 years and has its roots in military logistics research. Harman [14] distinguishes five different simulation types including real systems, replicate systems, lab-based simulation, where "a laboratory model might consist of the actual replication of some elements and the abstraction and substitution by symbolic representation of others", computer-based analytic and stochastic simulation, and analytical simulation. Lab-based experiments have the advantages of being economical, repeatable and requiring only a limited commitment [see, e.g., 14].

While the role of a logistics lab in research has been investigated in the 1950's, there has not been a deeper discussion on the topic ever since. Instead, other disciplines have further developed the concept of lab-based research. In engineering disciplines the role of labs in education has been investigated, for example, by Feisel and Rosa [15] and Leão et al. [16]. Baumgarten and Hildebrand [10] state that classroom teaching formats do not suffice in today's dynamic logistics education. Lab-based education is a valuable addition. Some research activities focus on collaborating, distributed research environments, such as Virtual Research Environments and Communities, which use computers and IT-networks for enhanced research across multiple disciplines and institutional boundaries [see, e.g., 17, 18]. Other researchers have focused on remote experimentation [see, e.g., 19, 20]. The boundaries between 'virtual' and 'remote' labs are not quite clear, though. Some researchers consider simulation to be the decisive part of virtual labs [21]. Virtual labs may use tools such as LabVIEW or Matlab/Simulink and are ideal for experiments that would otherwise be too costly, unsafe or not available in a real environment [21]. Additionally, animations and videos are used in virtual labs [16]. Simulations and pure virtual environments, however, cannot replace hands-on experiments in all fields of research as lab-based research is substantially different to theoretical research. Ragunath [22] claims that

- high marks in theory are of low relevance in the lab,
- experiment outcomes are usually unknown,
- experiments cannot be done "along the ways", but need focus, time and work.

Remote labs allow to access physical infrastructures at a reduced effort to train corresponding skills. In control engineering there has been a multitude of remote laboratory examples [23]. However, in logistics labs remote experimentation is not yet well established.

3 Components of a Logistics Lab

A typical logistics lab resembles production, warehousing and sales areas and utilises an inside space with different material handling devices, such as conveyors, storage systems and forklifts, to allow experiments in a realistic, yet simplified (e.g., shrinked), and controlled scenario. The boundaries to real environments may be seamless, as even off-campus industry sites may be used. The RFID Research Center at the University of Arkansas, for example, uses a warehouse area that is located in a company producing candles [24]. Additionally, the lab has access to a third party cold storage facility [24]. Smaller experiments may be conducted in office spaces. At universities, space limitations are a usual bottleneck to lab environments and competition for space is quite common. The inside area may be complemented by an outside area to test, for example, loading and unloading of trucks or locating of assets. Dock doors are used for the movement of goods between trucks and the warehouse area. For more controlled testing in relation to wireless technologies, anechoic chambers are used. However, as environmental impacts have a huge impact in logistics research "dirty laboratories" – a term that is commonly used to describe realistic interferences during testing similar to real environments – have certain advantages over anechoic chambers, especially for simulation of actual use case scenarios [24]. Logistics labs are expensive and require space and qualified personnel and "engineering department budgets are not always adequate to meet the needs of a modern instructional laboratory" [15]. Feisel and Rosa [15] see a conflict in lab financing for research over educational purposes. Thus, there is a need to work with tight budgets in logistics labs. Remote labs can help to optimise resource, space and time utilisation [16] and thus they will contribute to a more economic operation in logistics labs, if financial compensation between lab service providers and users can be achieved. Additionally, logistics labs allow research outsourcing for industry stakeholders and a seamless transfer of research results into business applications. However, logistics labs also offer an innovative sales and marketing channel – a trend that is obvious when seeing company logos in logistics labs and on corresponding websites – to promote new product developments for technology providers.

From a technology perspective, RFID has played a predominant role in logistics labs in the last decade. With commercialisation of the Auto-ID labs EPCnetwork development, EPCglobal and later GS1 has influenced the development of logistics labs considerably through an accreditation program for RFID Performance Test Centres [25] and the definition of several test methods [26]. However, these test methods are focused on retail demands and thus have limited relevance to other industries. In Europe, GS1 even has a tighter integration of labs – the European EPC Lab Network [27]. Other RFID testing procedures are provided by ISO [28] and VDI [29]. Testing methodologies are continuously adapted to the different fields of applications in logistics, thus comparison of results and replications of setups are difficult to achieve.

Computers have been used for simulation in logistics research from the beginning. The importance of computers for labs increased with the downsizing from mainframes to personal computers. Starting in the 1980s, special smart measuring devices with dedicated computer power enabled automatic data collection and analysis [15].

The Internet that developed from the ARPANET, a project of the Advanced Research Project Agency, enabled sharing scarce computing power (an issue that today has received new interest through GRID computing) and later remote access to physical lab infrastructures.

The importance of computer science as an integral part of logistics has further grown within the last decades through the introduction of EDIFACT, the concept of e-logistics [see, e.g., 30], the Internet of Things [see, e.g., 31] and recent Web 2.0 developments that focus on end-user integration and participation (not only) in supply networks. Consequently, holistic planning, control, coordination, implementation and control of information flows are considered in recent logistics definitions [see, e.g., 32]. The combination of information technology and logistics has lately been addressed through degree programs on 'Information Logistics' [see, e.g., 33,34]. In lab-based research this trend has been supported through the Auto-ID labs, which were established in 1999 to address research on the topic of the 'Internet of Things' – a vision that combines physical objects and Internet technology and utilizes RFID as a core technology. The development that originated in the Auto-ID labs lead to the EPCglobal Framework Architecture [35], which specifically addresses supply chain related communication and collaboration. The successful example of the Auto-ID labs that cooperated in research, supported by funding through GS1, unfortunately has not been multiplied yet. This may be blamed partially on competitive public funding schemes that usually focus on national or regional support. Logistics research, however, needs worldwide support to take into account the shift towards global supply networks.

The traditional 'technical' and 'organisational' view on logistics [36] should thus be complemented by an 'informational' view that focuses on automatic identification and measurements, data integration and aggregation, business intelligence and other information relevant topics in logistics. It may be argued that an informational view is a subset of the technical view. However, Delfmann, et al. [36] do not consider this in their 'technical' view of logistics that is restricted to material handling.

4 The Role of Business Economics and Engineering Disciplines in Logistics Research

The "instrumentation of logistics through 'hard' engineering technologies" [37] initially developed separately from business logistics. The focus of early logistic engineers was on productivity and scalability through automation in logistics and was later extended to material flow infrastructures, picking, packing, handling of returnable transport items, such as containers, and integration of computer-based planning and controlling [37]. The combination of engineering and economics in degree programs such as 'Engineering Management' supported the combination of engineering and business views in logistics research. However, the methods used in logistics research differ between engineering and business logistics. An overview about typical research methods in operations management, which is considered to be a sister discipline of logistics [38], is provided by Meredith, Raturi, Amoako-Gyampah, and

Kaplan [39]. The main methods used in lab-based logistics are prototyping, physical modelling, laboratory experimentation and simulation. Therefore, lab-based logistics research is described through a combination and "artificial reconstruction of object reality" and "rational, logistic positivistic/empiristic" research methods [39]. For other methods, logistics labs may be used for validation and demonstration purposes.

Combinations of these different methods, as described by Meredith, Raturi, Amoako-Gyampah, and Kaplan [39], are quite common in logistics. Therefore, Large and Stölze [40] clustered the different methods to analyse the usage frequency in logistics dissertations. Their four method clusters are: first *field and case studies*, second *prototypes and simulation*, third *pure normative or descriptive modelling*, and fourth *interviews, surveys and expert panels*. Prototypes and simulation are more common in engineering disciplines than in economics (cf. Fig. 1). Unfortunately, Large and Stölzle [40] do not distinguish *laboratory experimentation* and *physical modelling* from the framework of research methods in logistics and do not explicitly include these in their cluster analysis. 38% of the dissertations had their background in engineering whereas 62% were in the field of business economics [40].

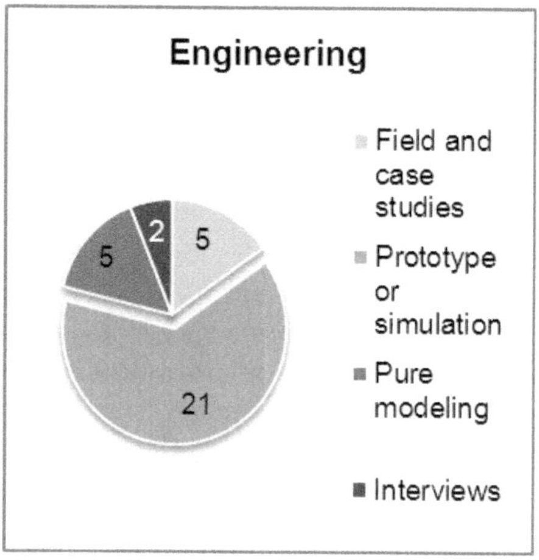

Fig. 1. Comparison of method clusters used in engineering disciplines [40]

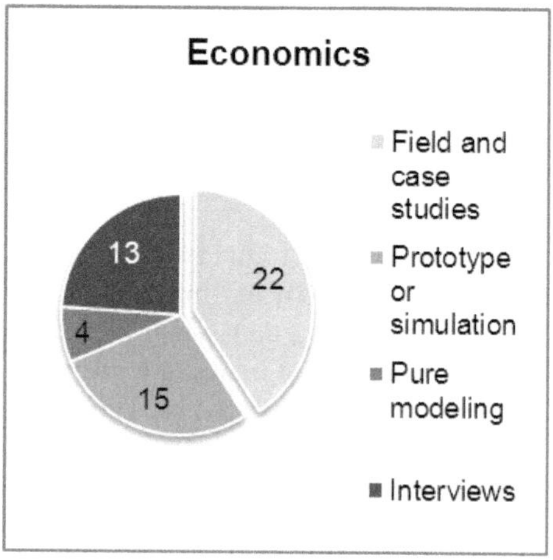

Fig. 2. Comparison of method clusters used in economic disciplines [40]

From the figures above it may be assumed that logistics labs play a more dominant role in engineering, because prototyping and simulation are the most used methods in this field. Therefore, investments in logistics research labs in engineering disciplines are more justifiable than in economic disciplines. Additionally, the operation of physical lab infrastructures is more common to engineering disciplines. However, non-engineering disciplines have a demand to access logistics lab infrastructures, for example, to verify their research approaches in a realistic environment.

Overall, there are 85 universities in Germany [41]. In a study by Baumgarten and Hildebrand [10] only 12 out of 34 German universities that offered logistics related studies and provided data on the orientation of their courses of study claim that they have a partial technology focus. Only 9 per cent of the universities mention that they have a direct technology focus (cf. Fig 3). This shows that technology and engineering related logistics studies are less represented than logistic studies in business economics. Nonetheless, missing access to logistics lab infrastructures may have impacted these results negatively. At the Universities of Applied science 19 per cent (more than double the amount of universities) claim that they have a technology focus (cf. Fig. 4). A review of university websites in Germany showed that for bachelor degrees Bachelor of Arts (43 %) is most common followed by a Bachelor of Science (41 %) and Bachelor of Engineering (7%). In master degrees, a Master of Science (57 %) is more dominant followed by Master of Arts (15 %) and Master of Engineering (7 %).

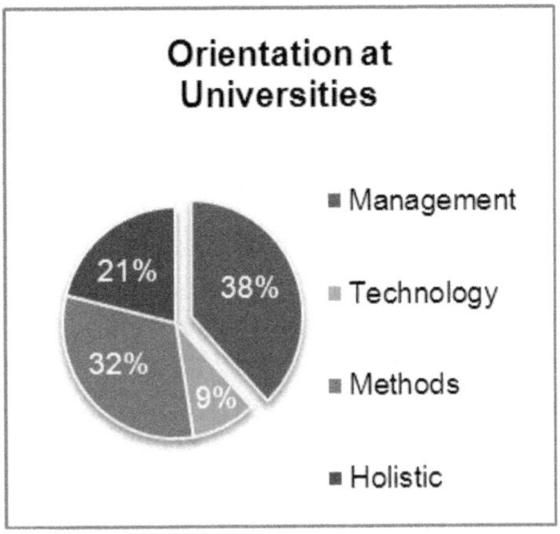

Fig. 3. Average orientation of logistics studies at 34 German Universities in 2008 [based on 10]

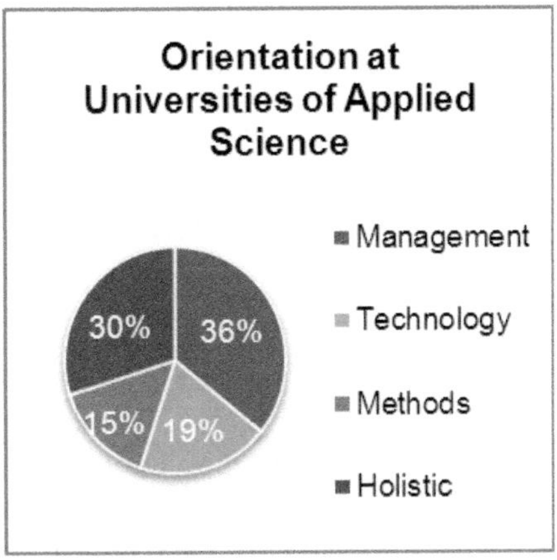

Fig. 4. Average orientation of logistics studies at 57 German Universities of Applied Science in 2008 [based on 10]

Unfortunately, Universities of Applied Science in general have less funding and, consequently, more restricted lab-resources. Infrastructure sharing between universities and Universities of Applied Sciences in Germany through remote experimentation would help to solve this. Even though there is a high number of Universities of Applied Sciences in Germany that offer logistics curricula, their relevance for scientific research is sometimes neglected. The non-profit German Logistics Association (BVL) in Germany, for example, has 15 professors in their scientific advisory council – none of which is from a University of Applied Science [42]. This seems surprising as logistics is seen as a field of applied science [see, e.g., 37] and the number of dedicated logistics courses of study at Universities of Applied Sciences outnumbers those at other universities [see, e.g., 10]. It has to be kept in mind though, that Universities of Applied Sciences focus on bachelor and partially on master degree programs, whereas PhD-programs are only possible in cooperation with universities. Consequently, average third-party funds per university professor (as one indicator for scientific contribution) are about 10 to 1 times higher (169.400 € vs. 17.200 € in 2007, [41]) than those of professors at Universities of Applied Sciences in Germany.

Logistics labs should be accessible to all institutions of higher logistics education, including universities, Universities of Applied Sciences, and colleges.

5 The Role of Logistics Labs for Future Logistics Research and the Need of Improved Collaboration

The current requirements concerning logistics labs have to be extended to meet future needs in logistics research. Ballou [3] identified nine major challenges for the future of logistics and supply chain research, which can be grouped into *benefit measurement and compensation*, *internal and external coalescence*, and *requirements for higher education*. The same structure can be applied for lab-based logistics research.

1. *Benefit measurements and compensation including revenue generation strategies and boundary-spanning metrics to identify, realise and share supply chain benefits:* The topic of benefit measurement and compensation in logistics can be researched in a lab environment. In the LogDynamics Lab at the University of Bremen this issue has been addressed through integration of information systems and the integration of a billing solution [43, 44, 45]. With this approach, it is possible to price and bill events (e.g., shipping, receiving), usage time (e.g., for returnable transport items), and information access (e.g., premium queries), thus enabling new revenue generating business models. While the concept could be proven in a lab-based scenario, a further evaluation in a real case scenario is needed. Additionally, the question of benefit measurement and compensation between institutions that maintain logistics lab infrastructures and those that access them (internal and external) needs to be solved.
2. *Internal and external coalescence:* Industry departments, such as operations, purchasing and logistics, are considered to merge organizationally under a supply chain concept [3]. Merging logistics researchers from different disciplines in a supply chain faculty may be necessary to mirror this development on a university level. Coordination, collaboration, and trust should be addressed in collaborative research environments, such as networked labs, to investigate the requirements in order to deduct findings for real company environments. Information sharing may be based on advanced information technology. A real lab network combined with virtual elements, such as simulation and the Internet of Things, can be used to create an appropriate research environment. In combination with time-limited, project-driven need for access to lab infrastructures, research findings from coordination of these multi-lab environments can help to understand similar processes of selective and short term supply chain relationships.
3. *Requirements for higher education:* As logistics curricula transform to supply chain curricula, relationship and trust building need to be further investigated.

Networked logistics labs can provide a realistic scenario in which these topics can be researched.

6 Conclusion

The relevance and impact of logistics labs for research and education should not be underestimated. The role of logistics labs should be reconsidered to meet current and future needs. The opportunities for lab-based logistics research, especially in networked environments, are manifold. Fig. 5 shows an overview of networked virtual and real logistics labs that allow local and distant access to multiple stakeholders.

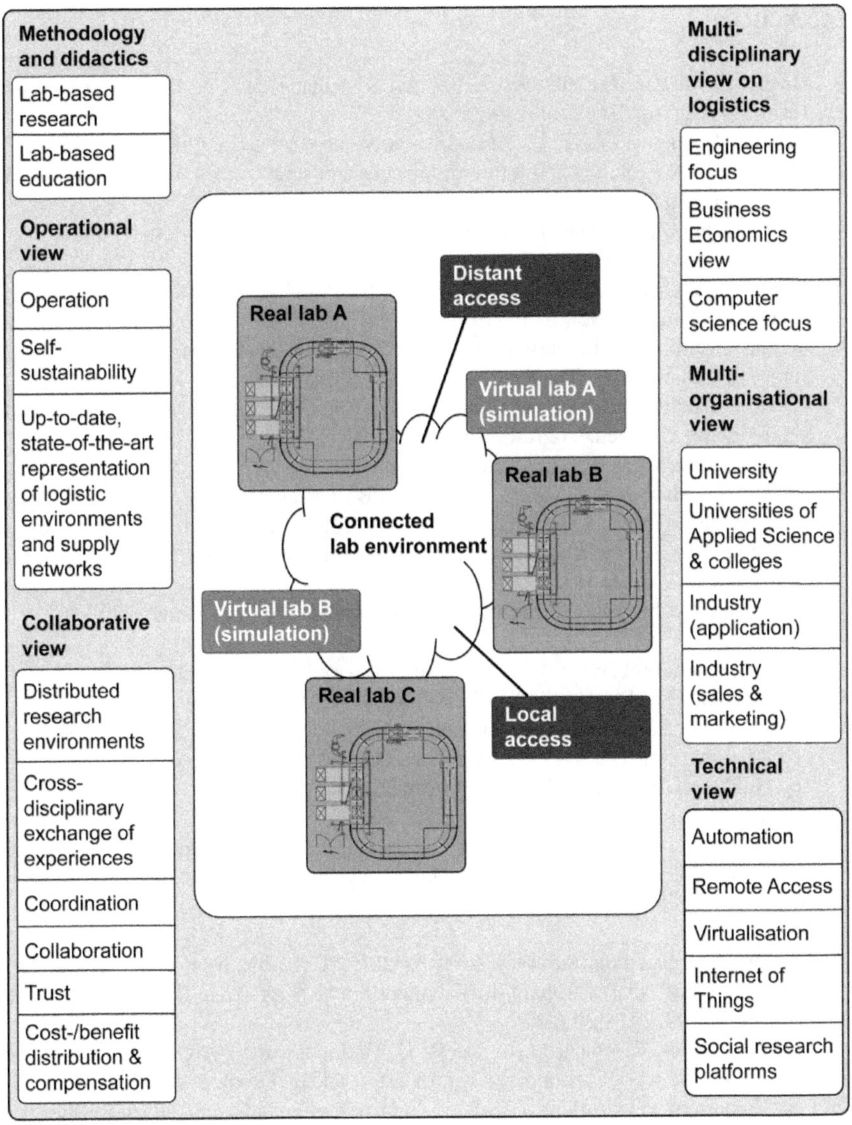

Fig. 5. Summary of different views on a networked logistics lab environment

Concepts of sharing information through the Internet of Things may be developed and demonstrated; trust mechanisms (or better contracts, auditing, and financial compensation) may be implemented and lead to a better understanding of supply networks.

The different views that have been covered in this paper may serve as a guideline for enhanced lab-based research and education in the future. Hopefully, funding institutions will address the need of international financing to enable the provided vision.

References

1. Mueller-Goldingen, C.: Zur Geschichte und Systematik des Begriffs Logistik. Wiss. Z. Tech. Univ. Dresd. 58(1-2), 17–19 (2009)
2. Ihde, G.: Trasport, Verkehr, Logistik, 3rd edn. Vahlen, München (2001)
3. Ballou, R.: The evolution and future of logistics and supply chain management. Eur. Bus. Rev. 19(4), 332–348 (2007)
4. Morgenstern, O.: Note on the formulation of the theory of logistics. Nav. Res. Logist. Q. 2(3), 129–136 (1955)
5. Smykay, E.W., Bowersox, D.J., Mossmann, F.H.: Physical Distribution Management. The Macmillan Company, New York (1961)
6. Heskett, J.L., Ivie, R.M., Glaskowsky, N.A.: Business Logistics: Management of Physical Supply and Distribution. The Ronald Press, New York (1964)
7. Material Handling & Logistics: Council of Logistics Management to become Council of Supply Chain Management Professionals, http://mhlnews.com/global/outlog_story_5987/
8. Arvis, J.-F., Mustra, M., Ojala, L., Shepherd, B., Saslavsky, D.: Connecting to Compete 2010: Trade Logistics in the Global Economy - The Logistics Performance Index and Its Indicators, http://siteresources.worldbank.org/INTTLF/Resources/LPI2010_for_web.pdf
9. Bundesverband Logistik: Wirtschaftsbereich Logistik 2011 mit voraussichtlich 222 Milliarden Euro Umsatz, http://www.bvl.de/presse/aktuelle-meldungen/tdl-pk
10. Baumgarten, H., Hildebrand, W.-C.: Studium Logistik: Akademische Ausbildung und Führungskräftenachwuchs in der Zukunftsbranche Logistik, http://www.logistik-berufe.de/downloads/Kap_04_Studie.pdf
11. Zeit Online: CHE Hochschulranking (December 2011), http://ranking.zeit.de/che2011/de/
12. Craighead, C., Hanna, J., Gibson, B., Meredith, J.: Research approaches in logistics: Trends and alternative future directions. Int. J. Logist. 18(1), 22–40 (2007)
13. Haythorn, W.: Simulation in Rand's logistics systems laboratory. Rand, Santa Monica (1957)
14. Harman, H.: Simulation: A Survey. In: IRE-AIEE-ACM 1961, New York, pp. 1–9 (1961)
15. Feisel, L., Rosa, A.: The Role of the Laboratory in Undergraduate Engineering Education. J. Eng. Educ. 94, 121–130 (2005)
16. Leão, C., Soares, F., Machado, J., Seabra, E., Rodrigues, H.: Design and Development of an Industrial Network Laboratory. Int. J. Emerg. Technol. Learn. 6, 21–26 (2011)
17. OSI: Report of the Working Group on Virtual Research Communities for the OSI e-Infrastructure Steering Group, http://www.nesc.ac.uk/documents/OSI/vrc.pdf

18. Fraser, M.: Virtual Research Environments programme: Phase 2 roadmap, http://www.grids.ac.uk/twiki/pub/EResearch/VreRoadmap2/JISC-BP-VRE2-final.pdf
19. Salzmann, C., Gillet, D., Latchman, H., Crisalle, O.: On-line engineering laboratories: real-time control over the Internet. In: The 106th ASEE Annual Conference and Exposition (1999)
20. Esche, S.: Remote Experimentation – One Building Block in Online Engineering Education. In: 2002 ASEE/SEFI/TUB International Colloquium on Global Changes in Engineering Education, Berlin (2002)
21. Chen, X., Song, G., Zhang, Y.: Virtual and Remote Laboratory Development: A Review. In: Song, G., Malla, R. (eds.) Earth and Space 2010: Engineering, Science, Construction, and Operations in Challenging Environments, pp. 3843–3852. ASCE, Reston (2010)
22. Raghunath, M.: Lab-based research is the best education, http://www.nus.edu.sg/ngs/Raghunath_on_Education.pdf
23. Dorin Popescu, C.: Remote vs. simulated, virtual or real-time automation laboratory. In: IEEE (ed.) 2009 IEEE International Conference on Automation and Logistics, Shenyang, pp. 1410–1415 (2009)
24. Patton, J., Hardgrave, B.: An Overview and Introduction to the RFID Research Center at the University of Arkansas. IEEE Commun. Mag. 45(4), 12–13 (2007)
25. GS1: EPCglobal Performance Test Center Accreditation Program, http://www.gs1.org/docs/epcglobal/implementation/EPCglobal_PTC_Accreditation.pdf
26. GS1: Approved Test Methods, http://www.epcglobalinc.org/apps/org/workgroup/subscriber/download.php/12875/uhfg2_1_0_9-TestMethod-ConveyorPortal_1_1_4-20060405.pdf
27. GS1: European EPC Lab Network, http://www.gs1.eu/?content=210#germany
28. ISO: ISO/IEC 18046-2:2011: Information technology - Radio frequency identification device performance test methods - Part 2: Test methods for interrogator performance, http://www.iso.org/iso/iso_catalogue/catalogue_tc/catalogue_detail.htm?csnumber=46151
29. VDI: Requirements to be met by transponder systems for use in the supply chain: Test methods to check the performance of transponder systems (RFID), http://www.vdi.de/uploads/tx_vdirili/pdf/1417468.pdf
30. Damen, J.T.W.: Customer controlled e-logistics. In: Smith, B., Kidd, P. (eds.) E-business: Key Issues Applications Technologies, pp. 508–514. IOS Press, Amsterdam (2000)
31. Uckelmann, D., Harrison, M., Michahelles, F.: An Architectural Approach towards the Future Internet of Things. In: Uckelmann, D., Michahelles, F., Harrison, M. (eds.) Architecting the Internet of Things, pp. 1–24. Springer, Berlin (2011)
32. Bundesvereinigung Logistik: Wissenschaftliche Definition der Logistik, http://www.bvl.de/logistik-scm/defintion/wissenschaftliche-definition-der-logistik
33. University of Applied Sciences Stuttgart: Information Logistics, http://www.hft-stuttgart.de/Studienbereiche/Vermessung/Bachelor-Informationslogistik?set_language=en&cl=en
34. Centre for Information Logistics: Information Logistics, http://www.informationslogistik.se/index.cfm?id=27&l=9

35. GS1: The EPCglobal architecture framework: Final version 1.3, http://www.epcglobalinc.org/standards/architecture/architecture_1_3-framework-20090319.pdf
36. Delfmann, W., Dangelmaier, W., Günthner, W., Klaus, P., Overmeyer, L., Rothengatter, W., Weber, J., Zentes, J.: Positionspapier zum Grundverständnis der Logistik als wissenschaftliche Disziplin, http://www.bvl.de/positionspapier/positionspapier-logistik
37. Klaus, P.: Logistics research: a 50 years' of ideas. Logist. Res. 1(1), 53–65 (2009)
38. Gammelgaard, B.: Schools in logistics research? A methodological framework for analysis of the discipline. Int. J. Phys. Distrib. Logist. Manag. 34(6), 479–491 (2004)
39. Meredith, J., Raturi, A., Amoako-Gyampah, K., Kaplan, B.: Alternative Research Paradigms in Operations. J. Oper. Manag. 8(4), 297–326 (1989)
40. Large, R., Stölzle, W.: Logistikforschung im Spiegel wissenschaftlicher Publikationen. Eine empirische Untersuchung auf der Basis betriebswirtschaftlicher und ingenieurwissenschaftlicher Dissertationen. In: Pfohl, H.-C. (ed.) Logistikforschung: Entwicklungszüge und Gestaltungsansätze, pp. 3–35. Erich Schmidt, Berlin (1999)
41. Wissenschaftsrat: Empfehlungen zur Rolle der Fachhochschulen im Hochschulsystem, http://www.wissenschaftsrat.de/download/archiv/10031-10.pdf
42. Bundesvereinigung Logistik: Wissenschaftlicher Beirat, http://www.bvl.de/wissenschaftlicherbeirat/wissenschaftlicher-beirat
43. Uckelmann, D.: The Value of RF-based Information. In: Haasis, H.-D., Kreowski, H.-J., Scholz-Reiter, B. (eds.) Dynamics in Logistics, LDIC 2007, pp. 183–197. Springer, Berlin (2008)
44. Uckelmann, M., Harrison, M.: Integrated billing mechanisms in the Internet of Things to support information sharing and enable new business opportunities. Int. J. RF Technol. Res. Appl. (2010)
45. Uckelmann, D., Hamann, T., Zschintzsch, M.: Performance increase and benefit compensation in supply chains by partial information sharing and billing based on identification of returnable transport items. Int. J. RF Technol. Res. Appl. 1(1), 23–43 (2009)

A Concept for a Flexible and Scalable Infrastructure for Remote Laboratories

Robustness in Remote Engineering Laboratories

Karsten Henke, Steffen Ostendorff, and Heinz-Dietrich Wuttke

Ilmenau University of Technology/Integrated Communication Systems Group, 98683 Ilmenau, Germany

{Karsten.Henke,Steffen.Ostendorff,Dieter.Wuttke}@tu-ilmenau.de

Abstract. This paper describes a universal remote lab infrastructure. This new extended infrastructure consists of three parts: a control unit for performing practical experiments with the physical systems, located in our remote lab, a bus protection unit to interface the remote lab bus and to protect it from misuse and damage, as well as a protection unit, which protects the physical systems against deliberate damage or accidentally wrong control commands. These units are for control, interface, surveillance and protection of the connected physical systems. The general concept and setup is presented in this paper, including the possibilities and limitations of such an approach. This concept is an extension of the remote lab architecture proposed in earlier papers.

Keywords: Control Engineering Education, Web-Based Education, Virtual and Remote Labs, Web-Based Design Tools, Distance Learning.

1 Introduction

Our Integrated Communication Systems Group at the Ilmenau University of Technology has many years of experience in integrated hard- and software systems and over 10 years of experience dealing with Internet-supported teaching in the field of digital system design [1, 2]. We have developed a new teaching concept, called "Living Pictures" [3] that we use in several phases of the learning process. Living Pictures are highly interactive Java applets that can be used for demonstration as well as for experimental purposes, and also serve as tools in certain steps of the design process of digital systems. To complete the learning outcomes by own experiences, the students have to pass a practical examination in a lab. A task during this examination is to design a digital control system that controls one of various physical systems, for instance a model of an elevator, a production cell or a positioning table, etc. The knowledge about digital control systems is taught in the following Bachelor and Master courses.

- Technical Informatics,
- Design of Digital Control Systems and
- Design and Validation of Complex Parallel Systems.

Dieter Uckelmann et al. (Eds.): ImViReLL 2012, CCIS 282, pp. 13–24, 2012.

The lecture *Technical Informatics* is held for Bachelor students in the first semester. It deals with the basics of Boolean algebra, combinational logic and simple sequential circuits. Students learn different functional specification techniques (e.g., logical equation, truth table, schematics and automaton-based techniques). This course is supported by "Living Pictures" as well as the Web-based usage of a remote prototyping board [4, 5].

Main topics in the lecture *Design of Digital Control Systems* are various minimization techniques for logical expressions, dynamic effects in combinational and sequential circuits and the design of digital control systems – mainly based on finite state machine (FSM) descriptions. This lecture is accompanied by a set of tools, which students can use in a 4-hour laboratory [6].

Students in Master level courses have the opportunity to deepen their knowledge in the lecture *Design and Validation of Complex Parallel Systems*. In this course, the students learn different methods and tool concepts to design and validate complex digital systems. For a better understanding, some accompanying practical designs are beneficial. Unfortunately, such designs are too complex for a single lecture or a 4-hour laboratory session. That is why we organize "project seminars" during a whole semester. The time for the project seminar is calculated with 120 hours – 4 credits.

For all students, hands on experiences are important to deepen their knowledge about topics they learned during lectures. At our university we offer a remote laboratory, which gives the students the possibility to work on real world systems without the need to stand in line at a lab or the need to take care of opening hours.

In our contribution, we would like to present three new electronic components for our new remote lab concept and the new possibilities they offer. The first is one protection unit for each physical system, the second a bus protection unit. And thirdly, additionally to the existing control units a microcontroller-based unit has been implemented for software oriented design implementations.

2 The Design Process

The design process of digital control systems usually consists of the conceptual formulation and the design of the control algorithm to finally achieve a validated control. Fig. 1 gives an overview of this process and describes the design phases in detail.

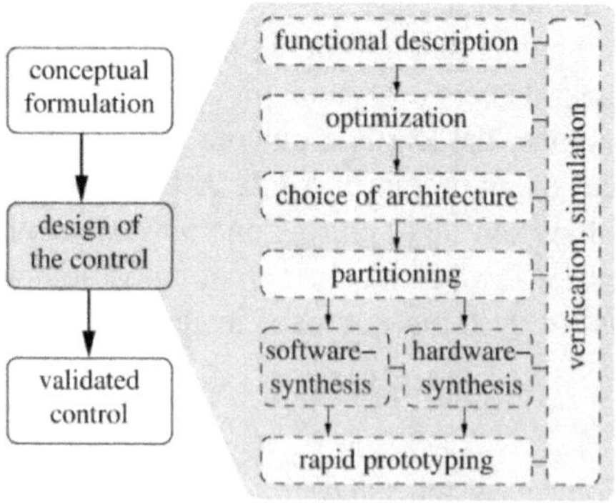

Fig. 1. Design process of digital control systems

For the *functional description* we use, for example, parallel finite state machines (FSM), Hardware Description Languages (AHDL, VHDL) or software notations (Assembler, C) as description techniques [7, 8]. These description methods allow a formal *verification* and *validation* of the design.

Simulation and visual prototyping help to find functional errors. Before starting practical work on real systems, simulations and animations in "virtual worlds" are often used to verify the developed solutions. The behaviour of the physical system that should be controlled, as well as its environment, will be emulated (e.g., as simulation model). The student can influence this "virtual world" and analyse the caused reaction of his control algorithm. Fig. 2 shows an example of such a visual model.

These steps have to be executed until no more errors are detected. But there is an essential disadvantage in this method. Real disruptive factors (e.g., failure of single components, mechanical problems or process variations) cannot be recognized by the underlying virtual environmental model.

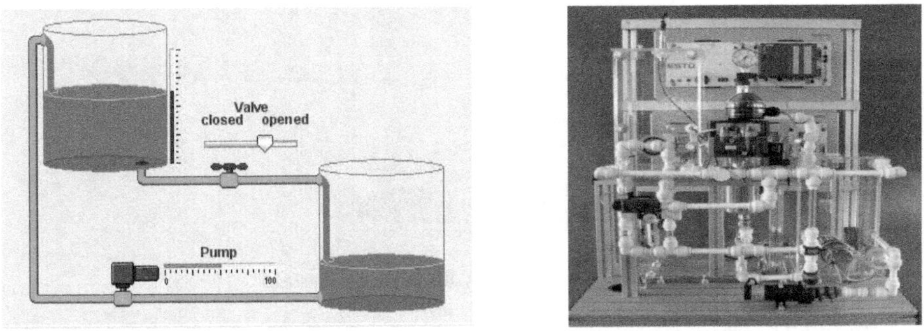

Fig. 2. Visual model and physical system of the water level control

Generally, only a simulation of predetermined malfunctions is possible. After some time, all these effects are well known in the student's community. Unconsidered sources of errors lead to undetected failures of the control because the corresponding environmental situation was not simulated before [9].

That is why a fault free design algorithm finally should be tested on real physical systems (e.g., the water level control, shown in Fig. 2) in the remote laboratory.

Assuming real laboratory conditions, disturbances can occur, causing special working conditions. Including such real disruptive factors to an algorithmic test leads, in contrary to simulation, to a distinctly closer relation to practical conditions. Furthermore, it should stimulate students' interest in the design of safety critical control systems. Additionally, we would like to give the students the chance to verify their prepared control algorithms and to correct or modify them accordingly to the received results.

3 Architecture of a Web-Based Rapid Prototyping Platform

An example for a tool set, supporting all the above mentioned design steps for complex control tasks, is the *REAL* system, developed within the Integrated Communication Systems Group at the Ilmenau University of Technology [10] – see Fig. 3.

As mentioned in further papers (for instance [11], [12], [13]), remote labs can open opportunities, allowing an experimental approach for a wider audience and an independence of opening times of the laboratory as well.

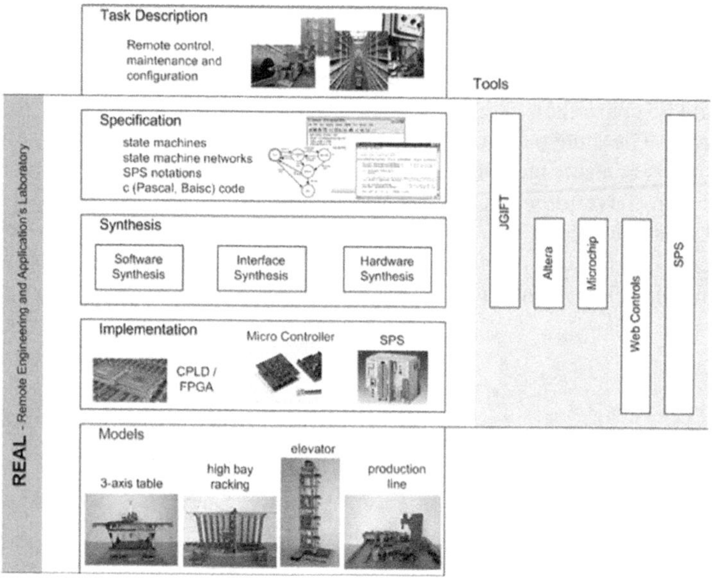

Fig. 3. Overview of the *REAL*[1] system

The *REAL* system offers various features. Visualization and animation allow observing and testing the properties of the design. In connection with formal design techniques, simulation and prototyping are used to establish a foundation for the development of a reliable system design. To check the functionality of the whole design, some special simulation and validation features are included as integral part of the *REAL* system. This offers various possibilities for the execution of simulations, such as:

- generation of executable visual prototypes from the general design,
- properties for step by step and parallel execution of these prototypes,
- visualization of the simulation process with the tools also used for specification,
- features for test pattern generation,
- code generation for hardware and software synthesis.

REAL offers a Web-based environment including verification and simulation features to generate and execute a design within visual prototypes (e.g., the visual prototype of the water level control, shown in Fig. 4).

As discussed in the beginning, it is furthermore necessary for students to test their design under real environmental conditions in the remote lab. For this purpose, we

---

1 Goal of the *REAL* (**Re**mote and **A**pplications **L**aboratory) system is to show new ways and chances of remote controlling and remote observation of real processes (e.g., in the fields of control engineering, robotics, tele-control engineering), dealing with integrated and interactive usage of modern Internet and intranet technologies, like WWW, HTML, Java, etc.

have developed a Web-interface to implement the control algorithm on the control unit. By using this Web-interface (as shown in Fig. 4), the student is able to:

- download the synthesized hardware and/or software control algorithms – generated and already validated by the *REAL* environment – to test his control algorithm on the physical system in the lab room,
- handle the experiment (start, stop, reset),
- change environmental variables if necessary and
- watch the experiment by manipulating environmental variables inside an I/O monitor or by observing the control of the physical system directly via a webcam.

At any time the students have the chance to adjust their algorithms in case of faults. Therefore, they are able to achieve a fault free solution (a validated control algorithm) step by step. For more details see [2], [9] and [14].

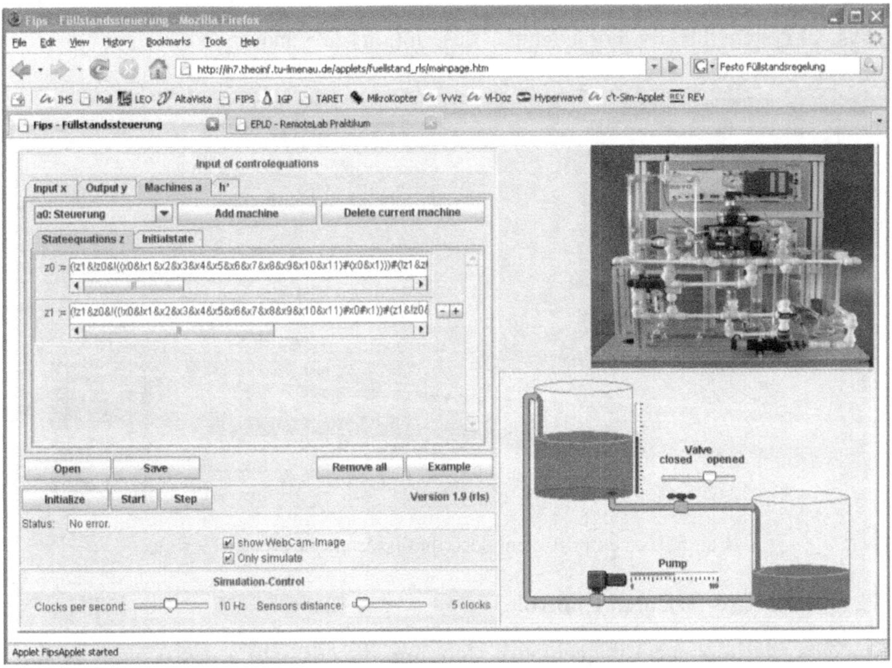

Fig. 4. Online regulation of the water level control with Internet connectivity to the remote lab

4 Fields of Application of the *REAL* System

As shown in Fig. 3, the *REAL* system supports the design of control algorithms with different specification techniques by an integration and usage of proprietary design environments as well as non-commercial software products. In the following, we will describe different specification techniques of the *REAL* system (FSM-interpreter-oriented, software-oriented and hardware-oriented) to design control algorithms for the physical systems in the lab room (for more details see [15]).

4.1 FSM-Oriented Control

In case of using finite state machines for specification, based upon an automaton graph, a student can use the JGIFT design environment [6] of the *REAL* system.

Assuming the student achieved a validated design, he gets the required next state and the output equations. By accessing the web browser interface of the *REAL* system, he is able to enter his algorithm (the received equations), handle the laboratory experiment (start, stop, reset) and change environmental variables if necessary. The control algorithm is executed by an interpreter – located either on the client or server side.

When using the "client-based" remote control, the physical system will be controlled "from a distance" through the interpreter running inside the client (implemented, e.g., as Java applet). In this case, only the input and output signals of the physical system will be transferred via Internet.

When executing the control algorithm "server-based", the required equations have to be transferred only once from the web browser to the interpreter running inside the control unit. In this case, the interpreter exchanges data with the I/O interface of the physical system directly. For a detailed description see 3 and 4.

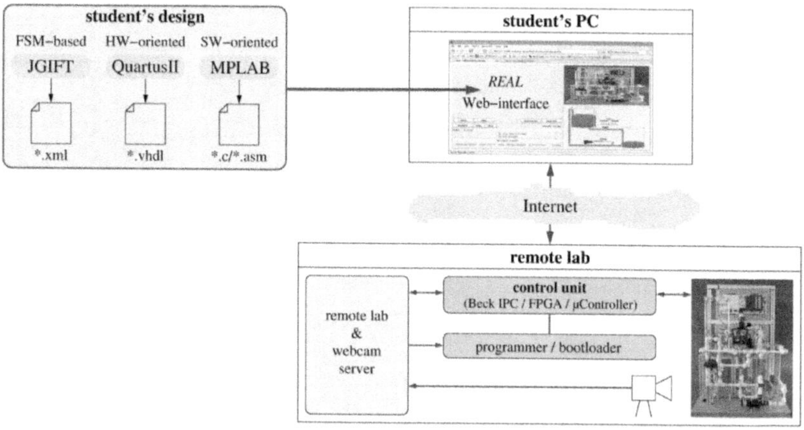

Fig. 5. Usage of different description techniques for the design

4.2 Software-Oriented Control

Students can implement their control algorithm directly into a microcontroller for a software-oriented implementation. Therefore, they use common (non-commercial) development tools, for example MPLAB IDE and/or C18 C-compiler from Microchip [16], to develop Assembler- and/or C-coded software projects.

The generated software control algorithm (Assembler- or C-code) is then transferred via *REAL* Web-interface to the remote lab server (see Fig. 5). The server automatically embeds the control algorithm into a locally stored workbench project which ensures safety of the physical system etc. After compilation, the project is programmed into the microcontroller, acting as control unit. Now, the student can begin with his experiment to proof, if his algorithm fulfils the requirements of the given task. A software implementation of the control algorithm is suitable for simple and non-time-critical requirements.

4.3 Hardware-Oriented Control

If a student prefers an exclusive hardware-oriented design using an FPGA (without any software components) and applying a hardware description language like VHDL as specification technique, he can prepare his design with common development tools, for example ISE, Max+Plus, Quartus II or others. The generated hardware-based control algorithm (e.g., a VHDL design) is uploaded via the *REAL* Web-interface to the remote lab server (see Fig. 5), which will synthesize the code for the FPGA using a locally stored workbench to ensure safety of the physical system. After configuring and programming the connected FPGA remotely, the FPGA is able to act as control unit for the designed control algorithm, and the student can start his experiment. For time-critical as well as parallel applications, a hardware implementation is the preferred solution.

5 Extensions of the *REAL* System

One implementation problem is to protect the physical systems in the lab against wrong control algorithms of unskilled students without defining too many constraints. Students should be free in their decisions and develop own creative solutions. They can implement their own design strategies. To protect the remote lab, a reference design and a method to check the students' design against this reference is needed [15]. The reference design should be independent of the used control unit and the development tools. This is done by the protection unit of the physical system.

This basic idea was the starting point for our extension of the *REAL* system and the design of a standalone protection unit for each physical system, which will be described in the next section (see Fig. 8).

In the current setup, every physical system in the remote lab is connected to one specific control unit (see Fig. 6). The user can connect to the selected control unit (e.g., a microcontroller, Beck IPC or an SPS) over the Internet. All functionality, including the reference model checking, has to be executed in the control unit. This imposes some constraints on the students' work:

- the reference design has to be included at design time,
- the student is forced to use a certain set of tools and project structure, and
- the correct use of the reference model in the compiled code has to be checked to prevent any intended or unintended manipulation.

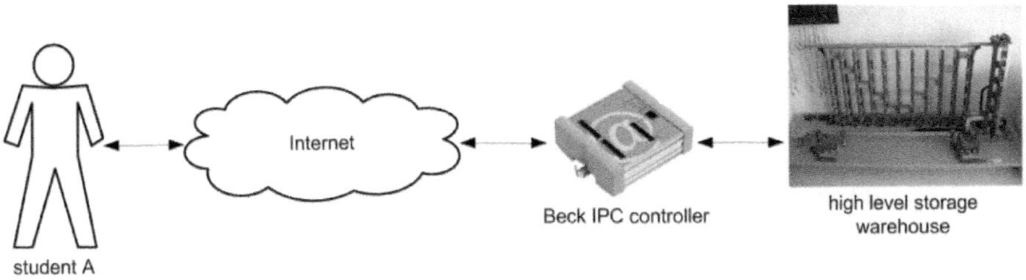

Fig. 6. Current setup of the remote lab

In our new approach, we want to eliminate a possible bypass of the protection functionality by taking it away from the control unit completely. This simplifies the development process of the control algorithm at the student side, because no special constraints apply. The student is free in using any control unit and development tool he likes. The architecture of the new concept is depicted in Fig. 7.

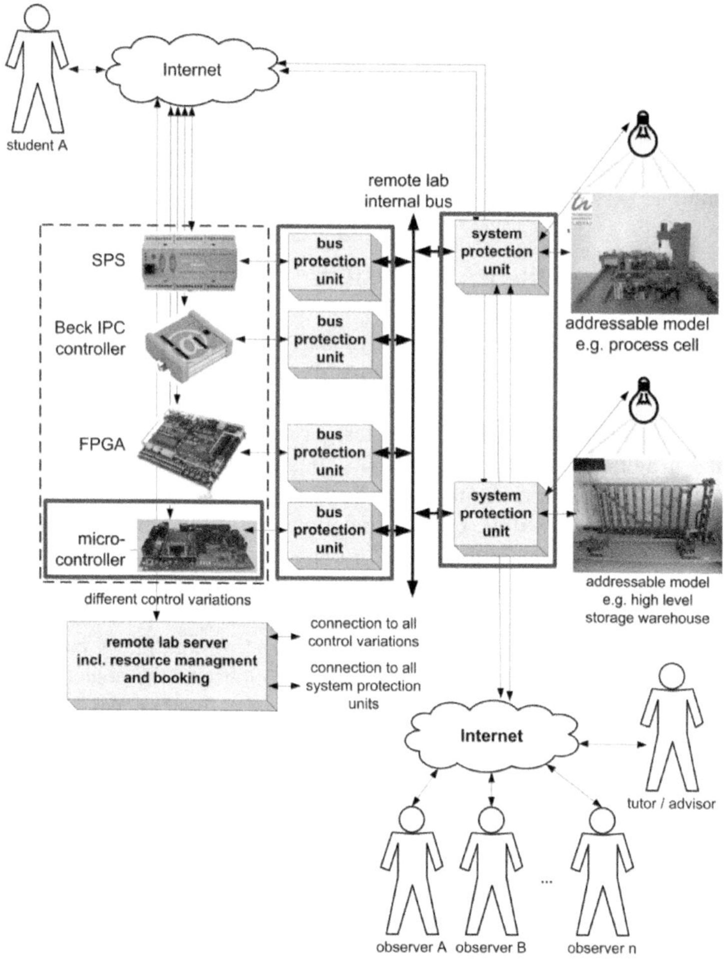

Fig. 7. Extension of the *REAL* system

Three new parts have been developed to make the remote lab more flexible, which will be described in this section. The first part, the *protection unit for the physical system*, is necessary when students want to be completely free in their choice of design tools and control units. The second part is a *bus protection unit*, which makes sure that, by accident or on purpose, no control device can transmit invalid bus commands or, even worse, just block the entire bus. The *microcontroller-based control unit* as the third new part of the lab will offer the students a new method for implementing their control algorithm.

5.1 Protection Unit for the Physical System

The protection unit is necessary when students execute their algorithms directly on the control unit and want to be free in their choice of design tools. Compared to the approach used so far, there was no possibility to check, if the executed commands are safe for the physical system. This means, damage could be caused by invalid commands. The task of the protection unit is to check for command safety by filtering all commands. Only commands that will not cause any malfunction are executed. All others are discarded and optionally reported as an error condition to a learning management system (LMS), see section 5.4 below.

This concept of a protection unit can be used with all control units like microcontroller, FPGA, SPS, etc. Using such a universal protection unit gives the students the largest degree of freedom for their design, because no precautions have to be taken into account. Therefore, no additional security framework (workbench) within the software and hardware control design is required to prevent malfunctions of the physical system (see section 4.2). The complete design flow is carried out at the students' side, giving them a more authentic look at a real world project design flow.

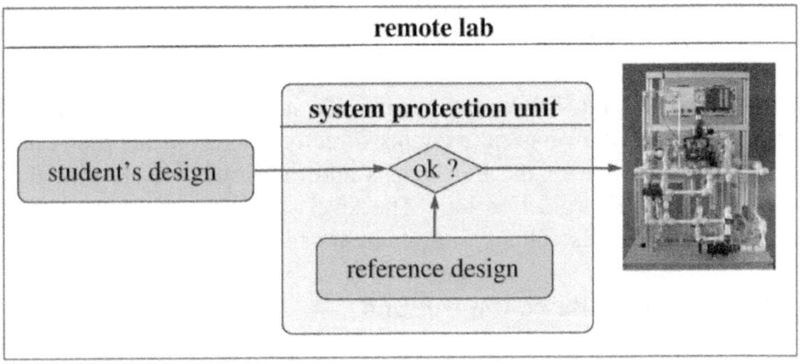

Fig. 8. Observation of the student's design by the protection unit

The central component of the protection unit is an XP2 FPGA from Lattice [17]. This FPGA is externally equipped with level shifters, to interface the physical systems, which have 24V logic levels. The FPGA can interface physical systems with up to 128 sensors/actors.

Besides the interface to the physical system, the protection unit offers a number of serial interfaces (USB, Ethernet) and an interface to the remote lab bus. In the preferred setup the protection unit is connected to a remote lab bus structure which allows a vast number of physical systems and control devices to be used in one lab. This reduces the complexity for interfacing many control units to one physical system and offers a large flexibility to the practical exercises that can be offered.

In addition to the physical system- and remote lab bus-interfaces, the protection unit has an Ethernet interface that connects it to the Internet. The interface consists of a hardware TCP/IP stack and a microcontroller. This interface is used by students for controlling some of the sensors on the e.g. when a button push should be simulated. Also, a number of observers and a tutor/advisor may use this interface to observe the physical system (see Fig. 7). The tutor/advisor may also control the physical system, for example to initialize some actors during an experiment.

The task of the protection unit to protect the physical system is carried out by running some state machines inside the FPGA. These state machines describe the possible safe behaviour of the physical system. This behaviour is compared against the commands received from the control unit to determine any unsafe control commands of a faulty student's control algorithm (see Fig. 8).

Commands that are valid for the current state of the physical system are passed. Commands that are not valid and could cause a failure or any other unwanted behaviour are discarded and optionally reported (see section 5.4).

5.2 Bus Protection Unit

The bus protection unit receives commands from a control unit and simply checks them for bus validity. This is done by verifying the transport protocol. The content of the transmitted data and addresses are not checked, because this is done by the specific protection units of the physical systems. The function of the bus protection unit is to prevent a control unit from blocking the bus and causing others to be affected. The bus protection unit is based on the same hardware as the protection units for the physical systems but uses a different population of components to simplify production and maintenance.

A secondary task for the bus protection unit is to interface control units like the SPS to the bus. This unit, for example, can only supply simple I/O control which is intended to interface the physical systems directly, therefore no protocol is implemented. The serialization of the I/O signals and the implementation of the protocol are handled by the bus protection unit. The level shifting, in case the control unit is not compliant with the bus voltage, is done by the bus protection unit as well.

5.3 Microcontroller-Based Control Unit

The microcontroller-based control unit comprises an 8-bit controller from Atmel [18], supports many connectors for extensions and gives the students the possibility to develop C- and Assembler-based control algorithms for a microcontroller.

In the area of the remote lab, the board also brings advantages over the old approach (see Fig. 6). This is mainly the boot-loader capability, which enables the controller to reprogram parts of its program memory by itself. Using a boot-loader in combination with the Ethernet interface eliminates the need for an external programmer, which otherwise must be connected to the remote lab server. Thus, the student can concentrate on the development of his C- and Assembler-based control algorithm and does not have to take the procedure of an external programming of the microcontroller into account.

5.4 Additional Features – LMS Interface

Besides the features already mentioned in this article, even more functionality can be added using the new concept of having a Web-based protection unit that checks the user input with a reference model.

The protection unit can be connected to a learning management system like "moodle" to forward any experimental results of the user. For an effective usage of the *REAL* system within learning management systems, the reference design and a method to check the student's design against this reference design step by step will be traced by the LMS. Fig. 9 shows this idea.

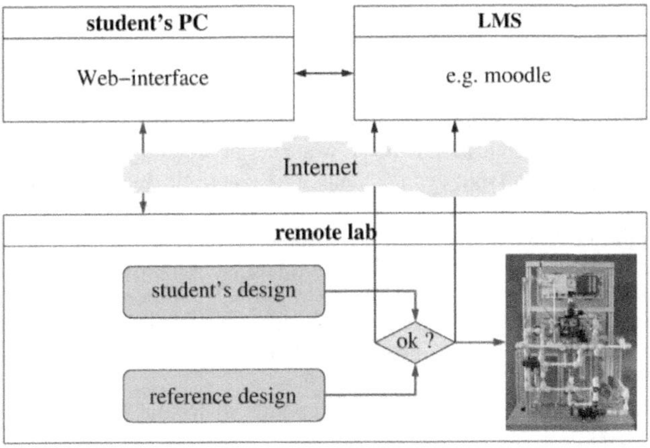

Fig. 9. Observation of the student's design under LMS control

If the student's design as well as the reference design produces the same output value, the resulting action can be executed by the remote lab. Otherwise, the LMS will be informed and the student gets some hints to correct his design. For a detailed description see references [15], [19] and [20].

6 Conclusion

We discussed an extension for our remote lab – a new microcontroller-based control unit for performing C- and Assembler-based practical experiments, a bus protection unit and a protection unit which protects the physical systems against deliberate damage or accidentally wrong control commands. These extensions offer possibilities for a flexible, scalable and robust usage of remote labs.

A connection to a learning management system is also possible. This can give feedback to the students for every experimental result taken by their implemented control algorithms.

Our Integrated Communication Systems Group at the Ilmenau University of Technology is involved in different national and international e-Learning projects [21, 22] in which it is increasingly necessary to allow and organize a shared use of equipment. That is why, the main focus of the *REAL* system is

- a Web-wide usage of different design tools and control units to control different physical systems in the lab room,
- a robust, fault-protected access to any connected physical system and
- an LMS-coupling for all control units and physical systems used in the remote lab.

These requirements can be fulfilled using the concept and the new components presented in this paper.

References

1. Sire, S., Geoffroy, F., Gillet, D.: A Virtual Assistant for Sending Hints and Perturbations to Students based on an Electronic Laboratory Journal (eJournal). In: Proceedings of the ITHET 2003, Marrakech (2003)
2. Henke, K., Wuttke, H.-D.: Web-based educational tool access. In: IASTED International Conference Computers and Advanced Technology in Education – CATE 2003, Rhodes (2003)
3. Wuttke, H.-D., Henke, K.: Living Pictures – tool-oriented learning modules and laboratory for teaching digital via internet. In: Proceedings of the ICEE – 2002 International Conference on Engineering Education UMIST, Manchester (2002)
4. Braune, T.: Rapid Prototyping components for Remote Engineering Applications (in German), Diploma Thesis, Ilmenau University of Technology (2007)
5. Ulbrich, S., Franke, S.: Case studies for a Rapid Prototyping Development board (in German), Study Thesis, Ilmenau University of Technology (2008)
6. JGIFT: Java-based Graphical Interactive FSM Tools, TU Ilmenau, `http://wcms1.rz.tu-ilmenau.de/fakia/index.php?id=780`
7. Hopcroft, E., Motwani, R., Ullman, J.D.: Introduction to Automata Theory. Addison-Wesley (2001)
8. Hamblen, J.O., Furman, M.D.: Rapid Prototyping of Digital Systems. Kluwer Academic Publishers (2002)
9. Henke, K., Wuttke, H.-D., Hellbach, S.: Laboratory via Internet – new ways in education and research. Int. J. Comput. Appl. 25, 157–163 (2003)
10. REAL: Remote Engineering and Applications Laboratory, `http://ih7.theoinf.tu-ilmenau.de/applets/index.htm`
11. Torroja, Y., et al.: A Modular Environment for Learning Digital Control Application. In: Microelectronics Education, Marcombo, pp. 185–188 (2002)
12. Fjeldly, T.A., Strandman, J.O., Berntzen, R., Shur, M.S.: Advanced Solutions for Laboratory Experiments over the Internet. In: Aung, W., Hicks, P., Scavarda, L., Roubicek, V., Wei, C.-H. (eds.) Engineering Education and Research – 2001: A Chroicle of Worldwide Innovations, pp. 135–146. Begell House Publishing (2002)
13. Wuttke, H.D., Henke, K., Ludwig, N.: Remote Labs versus Virtual Labs for Teaching Digital System Design. In: Proceedings of the Int. Conf. On Computer Systems and Technologies CompSysTech 2005, Varna (2005)
14. Henke, K., Wuttke, H.-D., Braune, T.: Virtual and remote labs in the Educational Process. In: International Conference on Remote Engineering and Virtual Instrumentation, REV 2007, Porto (2007)
15. Henke, K., Wuttke, H.-D., Braune, T.: Rapid Prototyping Modules for Remote Engineering Applications. In: International Conference on Remote Engineering and Virtual Instrumentation, REV2008, Düsseldorf (2008)
16. Microchip Corp., `http://www.microchip.com`
17. Lattice Semiconductor Corporation, `http://www.latticesemi.com`
18. Atmel Corporation, ATmega644A data sheet revision C, `http://www.atmel.de/AVR`
19. Wuttke, H.-D., Ubar, R., Henke, K., Jutman, A.: Assessment of Student's Design Results in E-Learning-Scenarios. In: 8th Conference on Information Technology Based Higher Education and Training (ITHET 2007), Kumamoto City (2007)
20. Henke, K.: Reusable Assessment Objects for Learning Management Systems. In: Computers and Advanced Technology in Education (CATE 2007), Beijing (2007)
21. Auer, M.E., Grout, I., Henke, K., Safaric, R., Ursutiu, D.: A Joint Master Program in Remote Engineering. Int. J. Online. Eng. 2(2) (2006)
22. TRE – International Summer School in Technologies for Remote Engineering, `http://www.fh-campuswien.ac.at/en/international/summer_school`

Virtual Logistics Lab: A Framework for Rapid Prototyping and Remote Experimentation

Carl Franz, Tobias Fritz, André Kreis, Maurice Meyer,
Maik Müller, Andreas Nawroth, Christoph Schröder,
Afshad Syed-Ahmed, and Igor Zhuchenko

Technologie-Zentrum Informatik und Informationstechnik,
Universität Bremen,
Am Fallturm 1, 28359 Bremen, Germany
grp-vllintern@informatik.uni-bremen.de

Abstract. In this paper the outcome of the student project Virtual Logistics Lab (VLL) is described. The open issues of already existing logistics labs are discussed and VLL's approaches to solve these issues are introduced. The dynamic and technical approaches are explained, including the flexible plug-in architecture for machines of warehouse logistics and the web-based user interface for remote control of the machine-complex at the LogDynamics Lab. The events of live demonstrations of VLL via XML based scenarios are described and options for future work are named.

Keywords: Virtual, Logistics, Lab, Web 2.0, Demonstrators, Remote, RFID.

1 Introduction

The VLL started in October 2009 as a project of 14 computer science students at the University of Bremen. The project was supervised by one professor and four research assistants. VLL was a cooperative project of the Research Department Artificial Intelligence located at the Center for Computing and Communication Technologies (TZI) and the LogDynamics Lab in Bremen, Germany. The duration of the project was two years.

New key technologies like Radio Frequency Identification (RFID) allow a more effective and efficient management of logistics. Before these new technologies can be incorporated in real-life production and assembly processes, they must be tested and analysed in laboratory environments. The problem was how to design an effective and flexible test environment and conduct experiments efficiently with little effort as well as location-independent.

The project aimed to create an integrated platform to control different devices via remote access over the Internet. A central control system for all the connected

Dieter Uckelmann et al. (Eds.): ImViReLL 2012, CCIS 282, pp. 25–34, 2012.

machines of a lab had to be developed. Another goal was the integration of multiple logistics labs. Accomplishing these goals would allow the machines of those labs to be used more efficiently and by international scientists.

2 Related Work

In VLL's context, two major research trends can be identified in the current development of labs. The first trend deals with RFID labs where specific components of the RFID technology can be tested. The second trend focuses on remote access to lab infrastructure.

A first exemplary RFID lab is located at the University of Parma, Italy[1]. The performance of various makes and models of the readers and antennas can be tested there. The lab's purpose is to determine how the hardware performs in real-life use cases. The testable hardware set contains a large spectrum of fixed and handheld readers, desktop interrogators, printers and tags. The manufacturers of RFID hardware can receive a certificate after a positive test result. For the tests, the lab has developed a set of approximately 10 use cases. [1]

The second lab is the AUTO-ID LABS Cambridge[2]. This lab does almost similar research on the impact of RFID. But here, the aspect of certification is less important than the focus on investigating the performance of UHF readers and tags in various environments. The goals of this lab are reducing the uncertainty of RFID measurements, finding methodologies for tracking and tracing objects, managing product information networks, quantifying the impact of RFID introduction, and integrating RFID with sensing and automation systems. There is also research on recommendations on time critical processes in applications that use RFID information. Both labs do RFID specific problem research that is conducted on-site by the lab carrier.

One of the first examples for remote labs is the application Automated Internet Measurement Lab (AIM-Lab) for remote experiments on semiconductors on the Web. It has been established at Rensselaer Polytechnic Institute (RPI)[3] and at the Norwegian University of Science and Technology (NTNU)[4]. The AIM-Lab provides remote measurement in the domain of semiconductors. The Software allows eight experiments to be performed on various semiconductors [2]. The measurement data is transferred to the client via server.

The second remote lab example is the Automatic Control Telelab. It is based on a Matlab/Simulink[5] environment. User-defined controllers can also be integrated. This lab is accessible for students at 24 hours per day. The Telelab allows the user to find machines that fulfil given requirements. There are four processes remotely available. One process contains the steps input, start, end and analysis, defined by the user.

---

[1] `http://www.rfidlab.unipr.it/en/`
[2] `http://www.autoidlabs.org.uk`
[3] `http://www.rpi.edu`
[4] `http://www.ntnu.edu`
[5] `http://www.matlab.com`

In the input step the user specifies the input values for the experiment. The step start runs the experiment and produces values. The user can observe the values and also the machine by webcam. At the end, these values can be analyzed by algorithms and the results are presented to the user. [3]

The third example is a remote lab developed at the Technical University of Catalonia. The software was created by using LabVIEW. The software enables the user to do various experiments. Every authorized student can get a timeslot to conduct experiments like the characterization of timed responses of thermal systems to different electrical stimuli, or identifying frequency responses of adjustable electrical filters. [4]

Another example for remote labs can be found at the University of Houston. It provides access to their remote laboratory for the Smart Materials and Structures Laboratory. They use parts of the LabVIEW software package[6] to control the hardware, develop graphical interfaces and publish the results on the Internet. For the Internet presentation ActiveX components are used. [5]

The most research on remote labs is supported by using technologies ranging from LabVIEW and Matlab/Simulink to Java applet, Flash, Ajax and other techniques. [6]

Open Issues. An open issue is that the RFID labs mentioned above are local, so the potential user has to be on-site with the machines to use the lab. The remote labs Chen reviewed in "Virtual and remote laboratory development: A Review" [6] have some more issues. These are missing compatibilities regarding plug-ins, used platforms and used operating systems. Another issue is that there is no standard framework so that developers from different disciplines cannot use that framework to develop their own experiments. Furthermore, there is the issue that many current remote labs use runtime engines that have to be installed by the user. The researcher has to know about computer hardware and software development to make his lab accessible.

3 Identification of User Needs

To identify the needs of potential users of a VLL and former users of the LogDynamics Lab, interviews were conducted. Therefore, a field manual was developed. The outcome of every interview was one or several so-called User Stories in a given format. A User Story consists of the role of the interview partner, the required function and an explanation, why the function is necessary in the lab. The requirements and goals for the VLL project were afterwards derived from these User Stories.

Important aspects formed from the User Stories were the requirements mentioned in the Dynamic Approach (see next chapter).

4 Dynamic Approach

Static structures are a major downside in a scientific environment where the configuration of an experiment changes frequently. For example, when a scientist wants to

[6] `http://www.ni.com/labview/`

run a modified or new experiment, which requires a new machine, the scientist has to be able to add that machine to the existing system. A fast modification would not be possible without rewriting a lot of the existing software. Rapid prototyping would not be possible with such systems but is needed to test new experiment configurations. Thus, a solution for this problem had to be found.

Another constraint of the viewed RFID labs was that they could only be used locally. Thus, a scientist who would want to use the lab would have to travel to its location. Not every scientist is willing to spend the time to travel only for running a few experiments. So a possibility to control the lab for a designated timeslot without being present in the lab would open all new options. One solution would be controlling the lab remotely via the Internet. Through this the utilization ratio could be improved, because scientists who do not have the financial means, time or the proper environment for building such a lab could use it as well.

In summary, the requirements for the new dynamic VLL are:

- Platform and operating system independency
- Universal central control of demonstrators using plug-ins
- Dynamic modification of the configuration of the running system for rapid prototyping
- Universal interface that makes special training unnecessary
- Integration of new demonstrators which can be implemented anywhere at runtime
- Global access and international use
- External presentation that lets others experience the technology
- Resource efficiency by full automation.

5 Technical Approach

To reach the desired level of flexibility, VLL is based on a dynamic plug-in system. It allows loading and unloading of device drivers at runtime. Every driver has access to the basic VLL functions like an event system and a database. The driver implements its functionalities as *Actions* that can be invoked by a VLL user directly or as a part of a larger scenario. These functionalities will be explained in this chapter.

5.1 Plug-In Framework

The plug-in framework allows loading of compatible plug-ins from a given URL. The plug-in does not have to be available locally and will be downloaded by the VLL. Every plug-in can control multiple machines of the supported type. For each machine an associated instance is created. To be compatible, a plug-in must implement its functions as *Actions* (see below), be multithreading safe and must not block the execution of other parts of the program (run asynchronously).

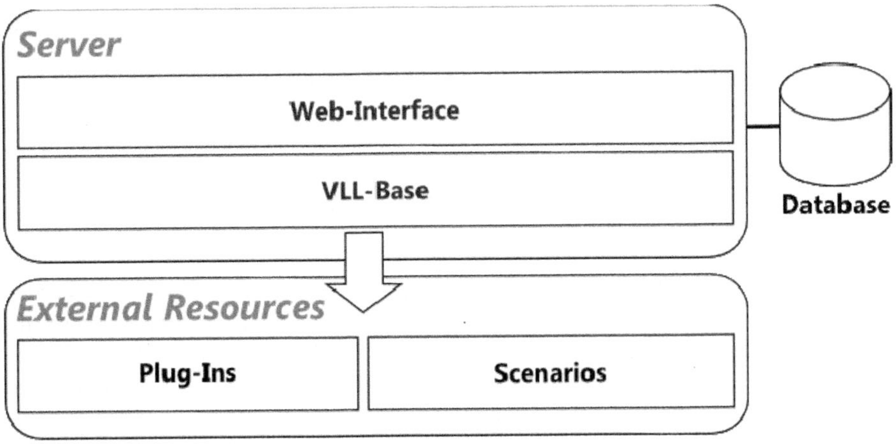

Fig. 1. Only the base VLL-Framework must be located on the server, all other parts can be loaded from other places on the Internet

5.2 Actions

An *Action* is an encapsulated functionality that can be triggered from the VLL core system or other plug-ins. A plug-in can register new *Actions* to the VLL to make functions accessible for the user. Triggering is possible manually through the web interface or automatically through a predefined *Scenario*. Usually, an *Action* triggers the corresponding function in the machine controlled by the related plug-in. *Actions* can use parameters and can return a success status or, in case of an error, a fail status.

5.3 Event System

As a plug-in must be able to communicate with every plug-in and the running scenario, an event system has been implemented. The event system is based on the Observer Design Pattern[7] and allows plug-ins to register and trigger events. For example, a plug-in can notify the scenario that an *Action* has finished.

5.4 Database System

VLL has a Hibernate session factory that grants plug-in access to a database. This has been used to track packets in the LogDynamics Lab (see next chapter) and store their position. Plug-ins can access the database to store data persistently. Also the plug-in connection data is stored in the database.

5.5 Scenarios

The *Scenarios* are the automation part of the VLL. A specific *Scenario* can be loaded, unloaded and controlled at runtime from a given URL. Scenarios are defined by an

[7] `http://www.research.ibm.com/designpatterns/example.htm`

XML-file and a corresponding background image. The image makes it possible to present the *Scenario* in a more user friendly way. The XML format is straightforward; it defines a directed graph of connected nodes. There can be different types of nodes. Currently, three types are implemented:

- *ActionNode*: This node triggers an *Action* and will wait until the *Action* has finished. This is the most useful node as it offers the main functionality.
  - *ActionNodes as condition checkers*: An *Action* can also check for a specific state in machine hardware or plug-in software. The node can therefore also be used to check conditions. The path of the *Scenario* can be chosen by determining if the node succeeded when the conditions were met or failed when they were not met.
  - *ActionNodes as sensors*: The event system allows the *ActionNode* to wait for an event. When, for example, a machine reaches a desired state, the node can succeed or fail at this point. By not deregistering the observed event, the node can be used as a permanent sensor while the *Scenario* is running.
- *WaitNode*: Simply waits a preconfigured time. Can be useful if a machine needs time to start up.
- *CountingNode*: Counts up and exits instantly on the first defined edge. If the preconfigured count limit is reached, the second defined edge of the graph will be called. Used in scenarios to iterate some repeating work, like "Discharge four packets from the high rack and after six round-trips stop the conveyor".

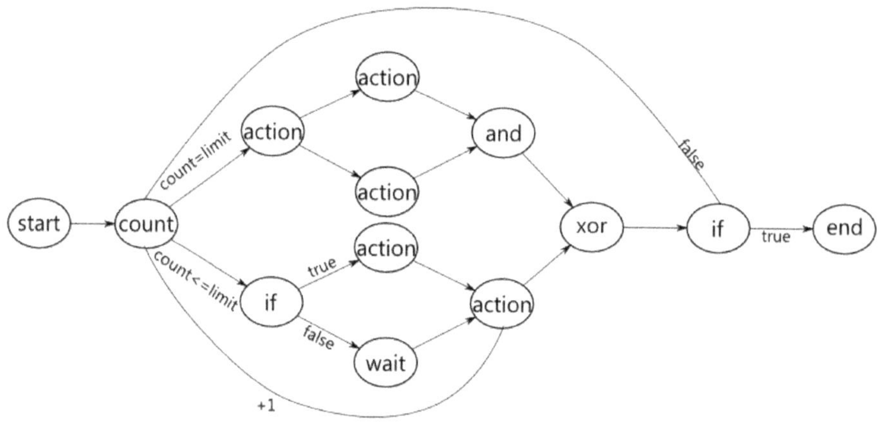

Fig. 2. An abstract *Scenario* featuring all possible nodes in an example setting

After one node is completed the next node is triggered. The graph can also branch out and run multiple branches (and therefore multiple nodes) at the same time. These branches can merge again, giving the node at the merge multiple preconditions to be triggered. The graph can also branch after a node fails, thus making error handling possible.

6 VLL in Use at the LogDynamics Lab

The VLL framework was installed in the LogDynamics Lab[8] at the Bremer Institut für Produktion und Logistik GmbH (BIBA)[9] in Bremen, Germany. "The LogDynamics Lab focuses on the optimisation of existing processes in production and logistics scenarios and the development of smart logistics products / services. It includes technologies to identify, locate, sense, communicate, process, and act." [7]

Plug-ins were written for most machines in the LogDynamics Lab. The machines consisted of components of internal warehouse logistics.

In detail the components (illustrated in Fig. 3) were:

- The high rack, that serves as a buffer storage for the VLL system. In the high rack, containers with various articles for the scenarios of the VLL are being held and can be stacked or discharged to the conveyor.
- The Reva-Server, which acts as a middleware for the networking of individual RFID readers that are located at the facility as well as for the process of reading tags and their further processing and therefore generating or processing of the Application Level Events (ALE). The system used at BIBA is a Reva Tag Acquisition Processor model 701.

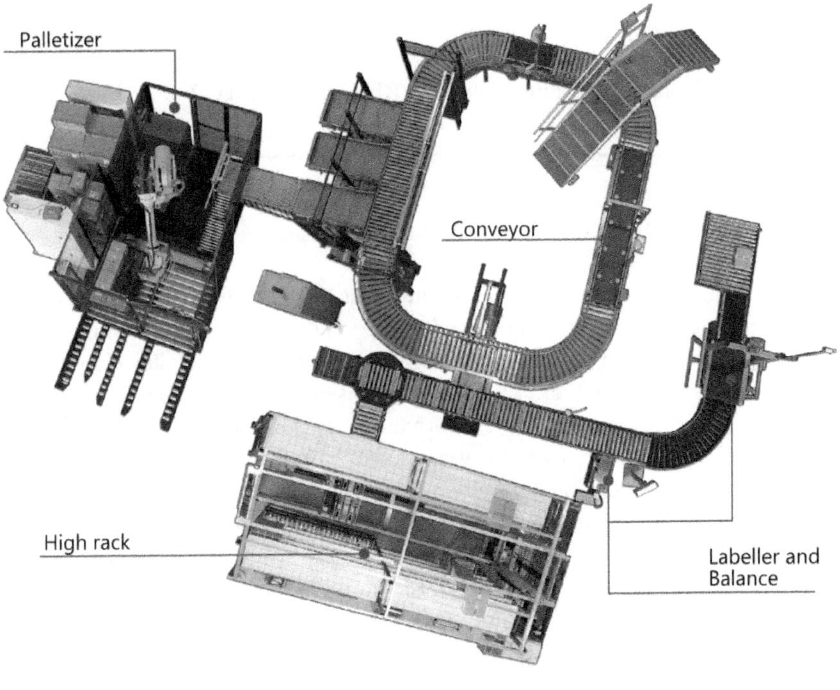

Fig. 3. An overview of the LogDynamics Lab with the positions of the used machines

[8] `http://logdynamics.biba.uni-bremen.de/lab.html?&L=1`
[9] `http://www.biba.uni-bremen.de`

- The Palletizer addresses incoming packets to a predefined spot on a pallet. This demonstrator is a palletizing robot of the company FANUC, model R-2000iB with the corresponding R-30iA control console.
- The pass through scales and labeller of the company Bizerba weigh and label passing packets with RFID labels containing the measured information.
- The conveyor is the central link between the high rack, palletizer and the scales and labeller with its function to control a PLC SIMATIC S7-300 from Siemens.

Table 1. Used machines with their functions and interfaces

Demonstrator	Function	Interface
High rack	Stacking, destacking, pick-by-light	TCP/IP: ORACLE
Conveyor	Conveying, change of speed, discharging, emergency stop	TCP/IP: Siemens CP 343-Lean (Fetch/Write)
RFID Server	Collecting and transmitting RFID tag information	TCP/IP: ALE Client
Palletizer	Palletizing, ejecting pallets	-
Labeller and Balance	Weighing, labelling RFID tags	TCP/IP

7 Presentation Events

After an implementation time of approximately one year, the first capabilities of VLL were presented at the *Internet Of Things 2010*[10] conference in Tokyo, Japan, as part of the demonstration of a prototype of the “Billing integrated EPCglobal Network” [8] of Bremen’s LogDynamics Lab. This system is based on an open source billing software and the Electronic Product Code Information Service (EPCIS).

A beverage supply chain *Scenario* was used to visualize the EPCIS-enabled synchronization of the flows of material, information and money. This approach allowed a flexible and scalable solution for cost benefit sharing and may enable new business models for the Internet of Things.

Also, after presenting the idea of the VLL on the *Supply Chain Day 2010*[11], one year later, an almost final VLL framework was presented at the *Supply Chain Day 2011*. A live demonstration of the capabilities of the VLL framework was shown.

[10] `http://www.iot2010.org/`

[11] `http://www.tag-der-logistik.de/en`

The capabilities of the VLL framework were demonstrated by running a *Scenario* live on-site. Several packets were discharged from the high rack to the conveyor. After running a couple of rounds, the packets were discharged to the palletizer and were processed.

8 Conclusion and Future Works

With the system that was created with the VLL project, it is now possible to integrate various machines easily with the use of a generic plug-in architecture. With the centralized web-based control mechanisms, users from around the world can conduct experiments on all machines that have already been connected to the VLL via a plug-in. There is no need to control the machines manually any more. Instead, the main functions of the machines can be controlled by the implemented actions in the related plug-in via the control panel of the system.

Thus, the daily work of the scientific personnel is eased and improved by the fact that machine control is now automated and centralized. Also, various experiments can easily be planned and conducted by the XML-based *Scenario* script system provided by the VLL. The scientists can control their experiments directly from their offices or even from other locations, due to the use of the Internet and several installed webcams.

For the most important problems VLL has found solutions, though not for all. The process of standardizing a framework for creating experiments has proceeded. Furthermore, the use of Web 2.0 technologies has been implemented. The problem of operating system compatibility was eliminated and there is no installed runtime required for end users.

The VLL has a great untapped potential for the future. Existing work can be used as a foundation for evolving and creating new functionalities. Recording and sharing experiments with other scientists could lead to a knowledge base with valuable results and may encourage more scientists to conduct their own experiments. This requires a user-friendly and self-explainable system to create experiments without prior knowledge. An easy to use *Scenario* editor with drag-and-drop for creating experiments could close this gap for users who do not have any experience with XML. [9]

The use of Multi-Agent Systems (MAS) in scenarios is another option to conduct experiments in new ways. Integrating MAS and therefore autonomous control by logistic objects could fundamentally change the way of creating experiments, as the user does not have to worry about how the output is processed. Using a user created *Scenario* input, an MAS would solve the problem on its own by making the right choices and communicating with the right demonstrators [10]. Some ideas for implementations of MAS technology in the VLL were brainstormed and prepared for future works like diploma theses.

As time always plays an important role in world wide used systems, experiments need to be done simultaneously in different labs. A job scheduling system could evenly distribute experiments in time slots or in different labs. This implies strong communication between labs and demonstrators and answers questions like how to save the state of a demonstrator, how timestamps are used in this communication and how the processes are presented to the user.

The VLL can be connected and integrated to the "Billing integrated EPCglobal Network" of Bremen's LogDynamics Lab as an inherent part and thus improve the billing functionalities of that system. [8]

References

1. Swedberg, C.: University of Parma's RFID Lab Launches Test Program for Apparel (2011), http://www.rfidjournal.com/article/view/8830/1
2. Shen, H., Xu, Z., Dalager, B., Kristiansen, V., Strom, O., Shur, M.S., Fjeldly, T.A., Lu, J.-Q., Ytterdal, T.: Conducting laboratory experiments over the Internet. Detection of Compositional Non-Uniformities in InP: Fe Via Spatially Resolved Photoluminescence and Secondary Ion Mass Spectrometry 31(3), 180–185 (1998)
3. Casini, M., Prattichizzo, D., Vicino, A.: The Automatic Control Telelab: a remote control engineering laboratory. Author Index 4, 3242–3247 (2001)
4. Anton, D., Bragos, R., Riu, P.J.: Remotely accessible laboratory for instrumentation and sensors. In: Instrumentation and Measurement Technology Conference, Como, Italy, pp. 1272–1276 (2004)
5. Song, G., Olmi, C., Bannerot, R.: Enhancing vibration and controls teachings with remote laboratory experiments. In: Proc. American Society for Engineering Education Annual Conference & Exposition, ASEE (2007)
6. Chen, X., Song, G., Zhang, Y.: Virtual and remote laboratory development: A Review. In: Proceedings of Earth and Space 2010, pp. 3843–3852 (2010)
7. LogDynamics Lab: Bremen Research Cluster for Dynamics in Logistics (2010), http://logdynamics.biba.uni-bremen.de/uploads/media/LogDynamics_Flyer.pdf
8. Uckelmann, D.: A Billing integrated EPCglobal Network - Synchronisation of material, information and financial flows. In: Internet of Things, IoT (2010)
9. Guinard, D., Trifa, V., Mattern, F., Wilde, E.: From the Internet of Things to the Web of Things: Resource Oriented Architecture and Best Practices. In: Harrison, M., Michahelles, F., Uckelmann, D. (eds.) Architecting the Internet of Things, Springer, Berlin (2011)
10. Schuldt, A.: Multiagent Coordination Enabling Autonomous Logistics. Springer, Heidelberg (2011)

A Living Lab for Internet of Things Vending Machines

Sauro Vicini, Alberto Sanna, and Sara Bellini

Fondazione Centro San Raffaele del Monte Tabor, via Olgettina 60, Milan, Italy
iris@eservices4life.org

Abstract. Vending machines are often considered mere dispenser facilities that trigger low engagement in their users. Instead, it is a market that is not only growing and expanding, but also evolving from a technological as well as service point of view. An experiential Internet of Things vending machine has been designed with the continuous involvement of users at different levels within San Raffaele Scientific Institute's City of the Future Living Lab in Milan. This paper illustrates the case of the Living Lab methodology adopted for the development of an innovative Internet of Things vending machine service.

Keywords: Living Lab, Internet of Things, Smart City, Service Design, Co-Creation, Vending Machines.

1 Introduction

This paper is explorative and induced from an ongoing development and practical implementation of a Living Lab for the City of the Future within San Raffaele Scientific Institute (HSR). It presents the case of an Internet of Things vending machine within a virtual as well as real research environment and community. Such setting is fruitful to explore and discuss how the Living Lab concept can be extended to the understanding, studying and measuring of the interaction dynamics among users and services offered, and the potential of Internet of Things technologies and their innovation.

1.1 Vending Machines: Market, Needs and Perspective

Vending machines are everywhere and are evermore so part of our lives because of our increasingly nomadic lifestyle [1]. There are about 3.7 million vending machines in Europe, run by some 10.000 companies, mostly SMEs and family businesses. Over 70% of vending machines in Europe serve hot drinks; the remaining 30% vend anything from sandwiches to cold drinks, hot meals and snacks [1]. This means that we are more and more forced to rely on outdoor sources of nutrition, which places a great deal of pressure on vending machine suppliers.

Indeed, the traditional model of goods distribution, known as driver sell, is often considered not cost-efficient enough and often leads to out-of-stocks [2]. This occurs because the driver acts as a type of salesman who has to visit the vending machines to

Dieter Uckelmann et al. (Eds.): ImViReLL 2012, CCIS 282, pp. 35–43, 2012.

refill the inventories as well as to collect money. This means that users visit vending machines without being aware of their current stock levels and there is no way of managing demands to avoid out-of-stocks.

There are also other important implications. Since consumers increasingly depend on vending machines for their feeding throughout the day (which can involve breakfast, mid morning snacks, lunch, afternoon snacks, and in some cases dinner), their needs are more articulated than in the past and the type, range and quality of products supplied must be adapted. Consumers are increasingly more careful about what they eat, they need to be able to buy their food quickly, and they are more technically adept (which means that vending machines are expected to communicate with 'things' and not only supply goods).

1.2 Internet of Things, Service Design and Smart Cities

The idea at the heart of Internet of Things is that all things and all environments can be improved from a functional point of view via the embedding of technology that remains invisible to the eye of the users, enabling both products and environments to become smart: meaning that they can gather data (or enable someone to gather data via them) from their surroundings, producing what Fleisch calls high-resolution data (or real-time data, essential in management and improvement of systems), as well as communicating amongst themselves as well as with humans enriching the value chain [3]. The Internet of Things is a strongly evolving field and it is useful to keep in mind that in the future it will be open, scalable, flexible, secure, customizable by its users and profoundly user-centric as stated by Uckelmann, Harrison and Michahelles [4].

Service Design is the activity of arranging and managing intangible goods, such as people, infrastructures, communication and products, for the attainment of users' goals and the fulfillment of their needs. The aim of this process, as expressed by the Service Design Network Manifesto, is to create services that are useful, useable, desirable, efficient and effective, based on a human-centered and holistic approach that focuses on the customer experience whilst integrating team-based interdisciplinary approaches and methods, in ever-learning cycles [5].

Internet of Things and Service Design are to be evermore intertwined. The service economy in developing countries is increasingly accountable for higher percentages of GDP and is mostly concentrated in financial services, health, and education. At the same time, products today have a higher service component than in previous decades, replacing the old dichotomy between product and service with a unified service-product continuum.

This is leading to greater expectations and more articulated demands both from users (for increasingly smarter and engaging products and environments) as well as by service providers and infrastructures (for more streamlined, real-time and differentiating artifacts). For this reason, the role of the Internet of Things is to become an integrated element of both services and their design process, for its capability of capturing, communicating and supplying data directly from the users and making it accessible to stakeholders and decision-makers.

Embedded technology in products and environments interconnected among one another in order to provide services through which users can achieve their objectives or satisfy their needs, is one way to interpret the concept of Smart Cities or Intelligent Cities. There is an array of definitions used to define what is meant by Smart Cities or Intelligent Cities in contemporary literature, but all of them stress the important role technology plays in improving communication, creativity, business, urban growth, social and relational capital, as well as social and environmental sustainability [6, 7].

In such a setting, vending machines are evolving and learning to embrace both Internet of Things as well as Service Design as ways to build an infrastructure for Smart or Intelligent Cities so as to respond more efficiently and effectively to users' changing needs and expectations.

2 HSR City of the Future Living Lab

San Raffaele Scientific Institute has recently set up the City of the Future Living Lab, both a virtual as well as real research environment and community. The unit is managed and organized by "e-Services for Life and Health" (http://www.eservices4life.org), a department of HSR specialized in the application of Information Technology to health, with the aim of developing and delivering services to the hospital's infrastructure as well as fostering innovation across numerous domains and disciplines. The Living Lab follows along the conceptual framework presented by ESoCE-Net, in which user-driven innovation is fully integrated within the co-creation process of new services, products and societal infrastructures [8].

The City of the Future Living Lab is an ecosystem where a multitude of stakeholders and partners can work alongside each other sharing knowledge whilst interacting with a wide variety of ICTs, therefore creating a fertile ground for innovation and cross-disciplinary research and communication. The City of the Future Living Lab is a miniature version of a city (with a hotel, a hospital, stores and offices, a supermarket and a post-office, a shuttle and bus service as well as streets and parks) and articulates itself in numerous scenarios. Its aim is understanding, studying and measuring the interaction dynamics among users and services offered and the potential of Internet of Things technologies.

2.1 HSR Living Lab Methodology

The research methodology implemented for the study of City of the Future scenarios is built on the widely recognized Living Lab process. The latter is based on four concurrent phases: co-creation, exploration, experimentation, and evaluation. Since this approach is an iterative and reflective one, a starting point is not defined and the Living Lab process can be commenced at any stage of the design activity. The figure below illustrates the Living Lab process adopted for building and evaluating the City of the Future services.

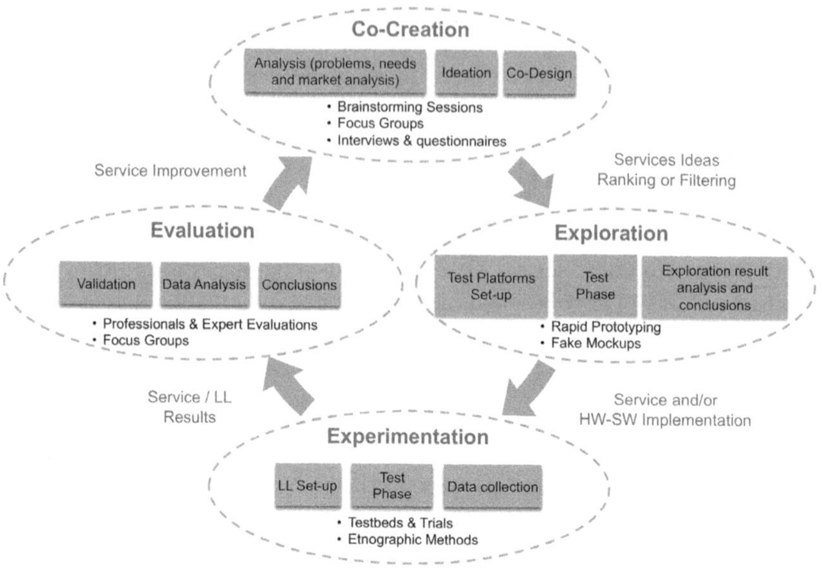

Fig. 1. Articulation of the Living Lab process adopted in HSR's City of the Future Living Lab

Living Labs are innovation environments that focus on user communities embedded within "real life" situations and environments. The fundamental concept at the base of a Living Labs is to gain direct and unfiltered access to users' ideas, experiences, and knowledge, based on their daily needs and desire of feeling supported by products, services, or applications. Users are directly involved in co-creating, exploring, experimenting and evaluating new ideas, concepts and technological artifacts related to Internet of Things applications and services. The Living Lab methodology is a design approach that focuses on making stakeholders and users constructive and active participants in the definition and construction of an artifact, be it a product, an interface, a service or an Internet of Things, with the aim of improving and building value into these same artifacts. Users and stakeholders are all involved from the early stage of the Living Lab process and throughout its entirety (therefore along the co-creation phase, the exploration phase, the experimentation phase, and finally the evaluation phase, as illustrated above in Fig. 1). Insights are gathered directly from the users in order to define and implement realistic, useful, desirable and effective artifacts.

2.2 Internet of Things Vending Machine Living Lab Co-creation

A number of different functionalities considered interesting to include as either technologies or part of the service were analyzed during the co-creation phase of our Internet-connected vending machine. The phase involved numerous stakeholders and possible end-users in brainstorming sessions, interviews and focus groups, and included vending machine producers, vendors and catering company managers, food company managers and other goods company managers, public transportation managers, nutritionists, psychologists, designers, computer and technologies producers, software developers, communication managers, personalized services experts, and policy makers. A myriad of necessities and possible implementable services emerged from these sessions, ranging from telemetry to user identification, from advertising possibilities to couponing and loyalty mechanisms.

Nevertheless, the most interesting and valid opportunity was that of a service for the promotion of healthy lifestyles, aimed at stimulating users to reflect upon the benefits related to healthy living through entertainment, motivation and socialization.

3 Internet of Things Vending Machine

The eServices for Life and Health unit has developed with its partners an innovative concept of vending machines (as illustrated below in Fig. 2) equipped with a touch screen monitor, CPU, Internet and wireless connectivity in order to meet the needs expressed during the co-creation phase. Indeed, these vending machines not only provide goods, but also the opportunity for its users to gather and to be entertained – when the machine is not in use, it operates as an entertainment tool displaying musical performances.

The vending machines feature an LCD touch screen panel in place of the usual glass front, through which all sorts of users (children through to mature users) can interact with the machine easily and comfortably. Users can view the selected product, its ingredients and nutritional details via the screen.

The vending machines are able to track all transactions and interactions, and record their usage history. Moreover, the vending machines are equipped with a webcam and a microphone that can register video and audio footage, which in turn provide information regarding the social interactions among users standing around them.

Fig. 2. The smart vending machine developed by the eServices for Life and Health unit

3.1 Telemetry

Telemetry is often adopted by vending machine operators to access real-time information from their vending machine network. Telemetry allows operators to access information, such as sales revenues, transaction data, as well as machine maintenance, and supply it in a remote and cost-effective manner, especially when machines are distant from one another or from the operator's headquarters [9].

Being able to identify a fault or error within a vending machine network (called Remote Monitoring) saves both time and money, for operators as well as for consumers. A telemetry system can immediately identify whether a coin or a product is stuck and can send out an alarm, so that a driver can intervene as quickly as possible to avoid losing customers, products or money.

Telemetry could also allow vending machines to communicate among each other, so that when the driver is in a specific area, he/she is made aware of the machines that are in need of maintenance or stocking up (called Curbside Polling). This allows drivers to optimize their routes, making the process of maintenance or replenishment fast, cost-effective and efficient. Both Curbside Polling as well as Remote Monitoring concurrently enable operators to keep their vending machines always stocked up, giving consumers more options to choose from and a more satisfying customer experience.

In response to consumers' needs for faster transactions, telemetry has also made it possible for them to pay quickly with direct debit cards and credit cards (called credit/debit acceptance) other than with cash. In the case where consumers are able to pay with cash, a so called Cash Management system notifies the operator when the machines require cash collection and refill, which increases consumers' purchase options and methods of payment.

3.2 Identification

The opportunity for vending machine users to be able to identify themselves during their user experience is the main way through which they can access a personalized service. At the moment, this can be made possible in the Milan area via a number of different digital tokens, such as the Lombardy Regional Card of Services, the Milanese travel card, and NFC Smartphones.

Through products such as smart cards, contactless cards, RFID and NFC devices, the user can receive contextualized and focused communication and marketing strategies as well as ad hoc interactions with the machine. This means that the user can receive personalized motivational information, entertainment material, couponing and loyalties, via the system's ability to match the data from the user's profiling according to what he/she has chosen to consume.

This, however, calls for a need of a third and neutral party above the users and the involved stakeholders (vendors and goods companies). The economical transaction guarantees the correct and ethical use of data, as well as the respect and secure management of the user's personal information.

3.3 Motivation, Entertainment and Socialization

As previously mentioned, the vending machine is not only a food dispenser but also a device for users' entertainment. Via its touch screen, speakers and microphone, the vending machine is able to interact with its users and provide the right information at the right moment. In this specific case, the vending machines are able to communicate information regarding the nutritional content of the food it offers as well as motivational information in order to stimulate and empower users to adopt healthy lifestyles. At the same time, these machines can also provide entertainment so as to add value to moments in which users interact with the machines, as well as stimulate users to interact among themselves and socialize, so that they are no longer mere facilities in non-places.

4 ELLIOT as a Tool for the Evaluation of Services

Since 2010, HSR has been participating to an FP7-ICT funded project via its e-services for Life and Health unit called ELLIOT. The ELLIOT project – where ELLIOT stands for Experiential Living Lab for the Internet of Things – aims to evaluate users' experience and attitudes towards several Internet of Things systems throughout all the four phases of the Living Lab process: co-creation, exploration, experimentation and evaluation.

A selection of four scenarios designed and created by the e-services for Life and Health unit has been identified to represent the City of the Future Living Lab within the ELLIOT project, and are complementary to the vending machine scenario illustrated up to now. The *Media Scenario* involves an interactive TV with a touch screen monitor and an easy-to-use interface suitable for children and is available in HSR's Pediatric unit. The *Tourism Service Scenario* is based on the Vainbici Web Portal (www.vainbici.it) and conceived for the promotion of initiatives for a healthier and environmentally friendly lifestyle. It also provides innovative services for the mobility of pedestrian-cycle users. The *Mobility InfoService Scenario* regards logistic services for all users of the automatic shuttle line of San Raffaele. The aim is to improve their mobility and to support patients, visitors, staff and students in their travel to and from the San Raffaele Hospital.

The goal of the ELLIOT project is to develop a software platform able to gather information regarding users' experiences whilst interacting with Internet of Things-based services. The ELLIOT experiential platform (ELLIOT EP) is expected to operate as gatherer and storage for the knowledge and experience obtained via the different Internet of Things. In particular, ELLIOT aims to measure the experience of users interacting with an Internet of Things service [10], either through embedded technology or by reproducing real-time data back to the research team.

Based on the data collected through ELLIOT, the services will be evaluated along the dimensions of a KSB experiential model. Each service will be assessed according to its ability to meet the Knowledge, Social and Business needs of the users. In the case of the vending machines, data will be collected via the embedded Internet of Things system (which at the moment includes a microphone, a video camera, an ID system, as well as the log data concerning transactions and goods' selection) that will subsequently be mapped across the KSB model to determine the vending machine's ability to meet the preconfigured objectives and outline areas for improvement and fine-tuning.

5 Conclusions

Living Labs are the most natural and effective means for the design, implementation, development, testing and validation of an Internet of Things system's pervasiveness. Activities of co-creation involving potential and final users is the most pragmatic, functional and economically convenient way to offer future users of the service something that truly responds to their needs and expectations.

HSR's City of the Future Living Lab is a real research environment and community that embodies a Smart City and is able to contextualize the object of a research in terms of service design and large-scale experimentation. A number of services have been designed and are being offered across different scenarios, with the objective of studying and measuring the interactions between users and services as well as the potential of Internet of Things technologies and their innovation.

Within the scenario of services related to innovative Internet of Things-oriented vending machines, the above methodology proves to be particularly relevant and effective, since traditional vending machines have always been treated as mere facilities. Vending machines' prices have been the main determinant in the past, whilst now their ability to offer innovative services that are both contextual and personalized provide an added value without precedents in the market segment.

The assessment of the user experience through the use of a Living Lab methodology that adopts technologically advanced solutions such as those present in the ELLIOT project is expected to dramatically increase the adoption of Internet of Things and to enhance the potential of collaborative innovation for the discovery of innovative Internet of Things application/service opportunities in bridging the technological distance with users/citizens.

References

1. European Vending Association: An Introduction to Vending (2007), http://www.vending-europe.eu/file-documents/information/Vending.pdf
2. Rusdiansyah, A., Tsao, D.: An integrated model of the periodic delivery problems for vending-machine supply chains (2005), http://prolog.univie.ac.at/teaching/LVAs/KFK-Seminar/SS06/11.pdf
3. Fleisch, E.: What is the Internet of Things? An Economic Perspective (2010), http://www.autoidlabs.org/uploads/media/AUTOIDLABS-WP-BIZAPP-53.pdf

4. Uckelmann, D., Harrison, M., Michahelles, F.: An Architectural Approach towards the Future Internet of Things. In: Uckelmann, D., Michahelles, F., Harrison, M. (eds.) Architecting the Internet of Things. Springer, Berlin (2011)
5. Service Design Network: Service Design Manifesto (2004), http://www.service-design-network.org/system/files/media/sdn%20Manifesto.pdf
6. Caragliu, A., Del Bo, C., Nijkamp, P.: Smart cities in Europe. In: Serie Research Memoranda 0048, VU University Amsterdam, Faculty of Economics, Business Administration and Econometrics (2009)
7. Komninos, N.: The Architecture of Intelligent Cities. In: Conference Proceedings from the 2nd IET International Conference on Intelligent Environments (2006)
8. Santoro, R., Conte, M.: Living Labs in Open Innovation Functional Regions (2010), http://www.esoce.net/Living%20Labs%20in%20Functional%20Regions%20-%20White%20Paper.pdf
9. Haddon, D.: Vending Operations: Reaping the Benefits of Telemetry Technology (2008), http://www.telemetrytalk.com/resources/MEI%20Authored%20Telemetry%20Article%208.12.08%20FINAL.pdf
10. Pallot, M.: D1.1 KSB Overall Framework and Experience Model (2011), http://www.elliot-project.eu/

Integrating Remote Laboratories in Post-secondary Science Education

Dietmar Karl Kennepohl[1], Jit Baran[2], and Ron Currie[2]

[1] Athabasca University, 1 University Drive, Athabasca, Alberta, Canada, T9S 3A3
dietmark@athabascau.ca
[2] Northern Alberta Institute of Technology, 11762 – 106 Street, Edmonton, Alberta, Canada, T5G 2R1
{jit,rcurrie}@nait.ab.ca

Abstract. The ability to provide the post-secondary student an excellent laboratory experience is crucial in many disciplines within the sciences and engineering. Remote access to teaching experiments offers educators another tool to incorporate a strong laboratory component with the added advantage of functionality and flexibility. This paper presents a comparative perspective to remote logistics laboratories by exploring remote environments for learning and research in the sciences. It is vital that the learning design of online teaching laboratories address the reinforcement of course principles and theory, as well as provide adequate student orientation to allow successful operation within a remote setting. The issues and considerations around integrating a remote experiment and creating a supportive student-centred learning environment are presented using our own experience with modern chemical analytical instruments as a concrete example. A brief analysis of this developing model is provided, as well as student and instructor feedback summarized as lessons learned.

Keywords: Distance Learning, Analytical Chemistry, Instrumental Methods, Remote Instrumentation, Student-Centred Learning.

1 Introduction

The laboratory is regarded as a fundamental part of the education and training for engineers and scientists and so university programs require practical components to be considered legitimate and recognized within the scientific community. Although we will use the general term "laboratory," it is equally applicable to other forms of applied learning components, such as design projects, clinical experience or field work. Traditionally, these laboratories have been face-to-face experiences. However, with the availability of various new information and communication technologies (ICTs) there are opportunities to supplement, if not replace, that in person laboratory environment. The two main ICT thrusts for laboratory work are the virtual laboratory (computer simulations) and, more recently, the remote laboratory (real-life but at a distance). The two terms are often confused with each other and used interchangeably. This is not

Dieter Uckelmann et al. (Eds.): ImViReLL 2012, CCIS 282, pp. 44–53, 2012.

surprising, because these new technology-reliant approaches to laboratories are also often found together and do strongly complement and reinforce one another. However, the discussion within this paper will be primarily on the latter.

Remote laboratories are increasingly finding their way into a variety of disciplines and quickly becoming a viable part of a science educator's teaching arsenal. One might ask why this has been such a relatively recent trend. Remote control devices are already well incorporated into our growing technological world and the concept of using remote control in the sciences is not new either. Scientists and engineers frequently exploit the advantages of remote access when the research they wish to conduct is physically inaccessible by virtue of location or danger. Although remote control has been with us for some time, remote control over the Internet for teaching experiments was only first established in the early 1990s [1; 2]. It was quickly picked up in such areas as robotics, computing, and engineering. In contrast, the use of remote access for teaching laboratories in the natural and physical sciences has not been as common. A detailed review of how this developed and the types of approaches employed can be found in [3]. There are several parallels in remote learning and research/collaboration environments found between logistic processes and networks and the sciences. Common issues include sharing equipment and instrumentation, creating appropriate remote learning and collaborative environments, engaging students, offering more flexibility and independence, and reflecting and using skills needed in the 21st century. Herein, we report on the effective integration of remote laboratories as a viable vehicle for teaching, learning and research collaboration in the sciences using the Canadian Remote Sciences Laboratories (CRSL) as a tangible example.

2 Role of the Laboratory and Its Pedagogical Underpinnings

The design of any laboratory component is often done to meet a variety of aims. The most general aim is the reinforcement of course concepts through illustration and making it real for the student. This is followed by the development of techniques and skills in the student that are either cognitive or practical in nature. It could include skills such as physical manipulations, observations, problem solving, planning, creation, data handling, interpretation of results, time management, and dealing with errors, to name just a few. In addition, many teachers cite the laboratory environment as an opportunity or vehicle to inspire students and make the course content and theoretical principles impressive enough to remember.

A tremendous amount of practical knowledge has been built up over the years around integrating the face-to-face laboratory component as part of the undergraduate course experience. No longer satisfied with just being content experts, there is also a growing interest among science educators in the scholarship of learning. Bailey and Garratt [4] provide a practical down to earth review of teaching and learning theories from a chemical education perspective, which can readily be applied to other science and engineering disciplines. Despite this substantial body of knowledge, there is comparatively little literature available on remote laboratories. Indeed, the literature on remote practicals and laboratories has mainly focused on the technology and feasibility of the access rather than on pedagogy or its educational value. Presumably, the remote laboratory would employ many of the same pedagogical strategies and learning designs as traditional supervised laboratories, including matching of teaching and learning styles, sufficient student engagement, incorporating appropriate feedback, and providing sufficient logical structure and guidance to form a cognitive foundation for the student to build on. Would it then not also have a similar instructional value as the traditional laboratory? Much more of this needs to be investigated and clearly articulated, because in universities non-traditional approaches to teaching are of interest but tend to be suspect. Not only are the teaching and learning cultures slow to change in academic organizations, there is also a great deal of value placed on direct human contact. Quite frankly, the real question on everyone's mind is whether there is equivalent learning occurring in the laboratory accessed remotely compared with the traditional on campus laboratory.

3 Learning in Remote Laboratories

There are problems commonly cited by critics around whether or not to incorporate a remote laboratory component within a course. First and foremost, it is not necessarily viewed to be a real laboratory experience. In part, this has to do with the conflation of virtual laboratories with remote laboratories. Unlike a virtual laboratory, which also has an important role in teaching and learning, the remote laboratory is not a simulation. It allows learners to physically carry out real experiments over the Web. Students obtain real results using real substances and make real conclusions, just as they would if they were in the laboratory with the equipment. Even with this clarification, the perception is that remote laboratories are not seen to offer the identical laboratory environment with all its atmosphere, noises, smells, and haptic experience of experimenting. Second, the student-student and student-instructor interactions are usually altered or reduced in a remote laboratory. The underlying epistemological assumptions of these criticisms are that both laboratory environment and direct human interaction are needed to give the student an effective laboratory experience.

Indeed, the laboratory environment and human interaction can and do lead to both formal and informal learning. However, we know that other forms of interaction in other environments can also lead to learning. Our own experience has been that there are no substantial overall differences in student performance between proximal and remote modes. This is echoed in the literature by other science educators who have also reported similar findings [5-8]. While overall learning appears to be equivalent, some researchers also report that proximal and remote modes each have their own strengths with respect to different components of student engagement and learning [9]. A knowledgeable teacher would therefore be able to exploit these to produce a quality laboratory experience. Another learning opportunity that should be considered here is the remote laboratory environment itself. As previously noted, practicing scientists and engineers are increasingly using remote access, so having exposure to this in a teaching laboratory is, in itself, valuable.

4 Designing a Remote Teaching Laboratory

Remote teaching laboratories are commonly employed in four basic ways:

1. to allow observations of natural phenomenon or experiments;
2. to carry out measurements;
3. to manipulate instruments or physical objects in experiments;
4. to facilitate collaborative work at a distance.

An individual experiment can contain any combination of these. The two major components to consider in developing an experiment are functional design and learning design. On the functional side, the experiment itself should be kept as simple and robust as possible, proper physical space needs to be secured to house the experiment, appropriate technology needs to be selected, ongoing maintenance needs to be arranged, and certainly ongoing financing should be in place. As with any laboratory component, there should be a clear vision of what learning outcomes are intended for the remote experiment being developed. Having the didactic foundation laid out is desirable before moving to the technology solutions needed to achieve those intended goals. In both functional design and learning design there are numerous approaches which have been described in [10-12], including some more creative approaches involving mobile devices, described in [13] and gaming, described in [14].

In addition to building the technological means to carry out remote experiments or teaching the disciplinary principles through a remote laboratory, the student also needs to learn how to be comfortable with and effectively operate in that remote laboratory environment. Successful remote laboratories are self-contained, intuitive, and designed with a seamless pedagogical front end to facilitate the high level of student learning and skills development necessary to carry out an experiment at a distance.

5 Canadian Remote Sciences Laboratories (CRSL)

The Northern Alberta Institute of Technology (NAIT) and Athabasca University (AU) have been developing teaching laboratories based on the control of analytical instruments in real-time via an Internet connection. The bulk of the work centres on analytical instruments in chemistry, but has also included one early experiment in first-year physics, namely a remotely controlled simple ball-drop apparatus used by physics students to determine the gravitational constant g [15]. Most modern analytical instruments in chemistry are intimately controlled by computer. They are extremely sophisticated instruments, but this also makes access by remote control possible if not facile. In an early pilot study, which allowed chemistry students to control analytical instruments in real-time, a series of 15 different analytical instruments were interfaced to computers, which in turn were connected in a local area network (LAN) in a laboratory environment. A server was then set-up to permit access to this LAN over a firewall using terminal emulation software (see Fig. 1). The student or guest accessing any of the instruments requires only a java enabled Internet browser at their location. Details of this initial setup as well as student performance and perspectives have been published in [11].

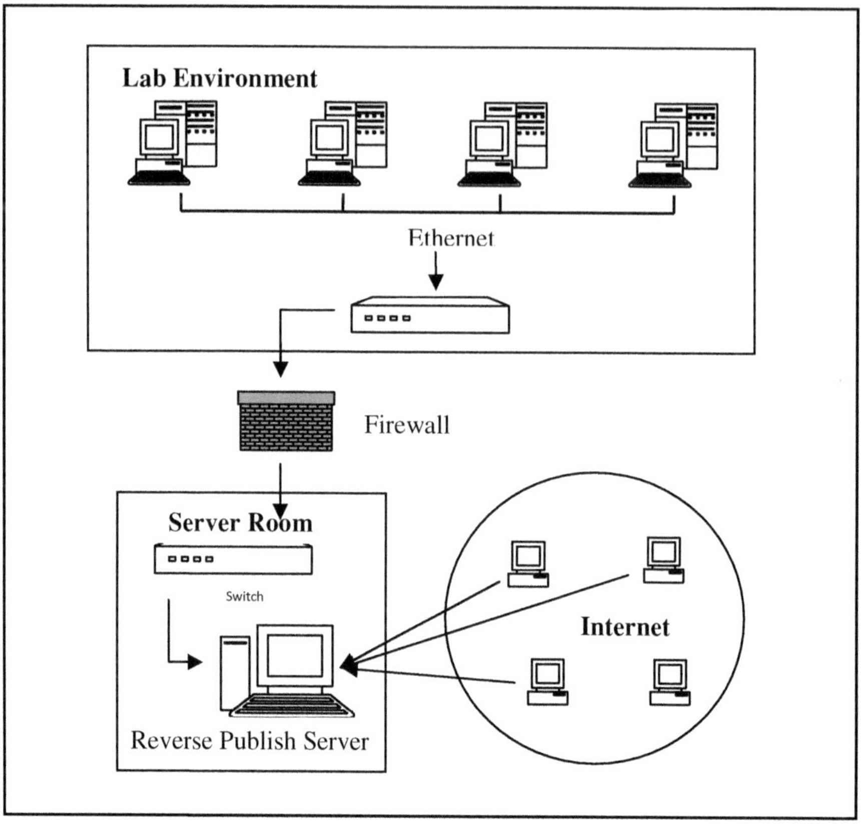

Fig. 1. Remote Access to Analytical Chemistry Instruments

Experiments in the remote laboratories were carried out by students at NAIT and AU as part of regular coursework. However, the system eventually ran into several technical difficulties that interrupted access to the instrumentation. In 2009-2010, a team of NAIT students, as part of their Bachelor of Technology capstone project (BTE 495), returned functionality of the CRSL system while better aligning it with current design practices and technological advances. They also established a flexible and sustainable system that should allow for ease of modification and scalability. In the new model, the software architecture had changed somewhat and included features like web services and a business logic layer to provide a simple method for programmers to implement any needed business rules requested by the sponsor group. The remote desktop interface initially employed by the students was a combination of TightVNC and vncviewer, which are both open source software and allow for easy web browser access. However, due to institutional security protocols, we eventually went to VMWare, which also allowed access external to the institution. Other updates were also made including installation of new web cameras, and an updated scheduler (phpSheduleit) to permit users to book time on a particular instrument, to name a few. Some features like the e-logbook and chat features had not been re-established, but efforts are underway to for future implementation. Together the work represents a reconnection of the remote laboratories in CRSL as well as an update/overhaul of the system itself (see Fig. 2). It is worthwhile to note that using the capstone project route with a team of senior students having backgrounds in computer systems technology, chemical technology and electronics engineering technology was a creative approach in remote laboratories and science education integration.

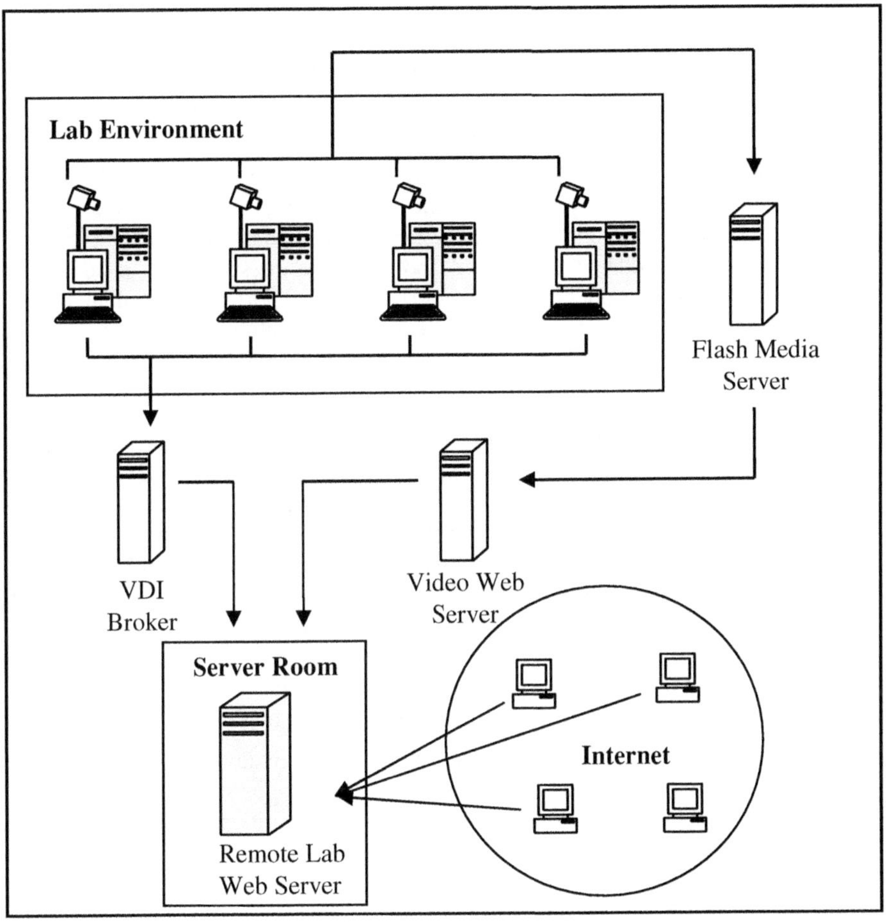

Fig. 2. Updated Remote Access with VMWare Solution

6 Incorporating the Remote Laboratory

Having established (or re-established) an updated remote access to the analytical instrumentation is only a first step. Building a seamless pedagogical front-end for instrument access in order to facilitate the high level of student learning and skills-development necessary to carry out an experiment at a distance is critical. The challenge for us as chemical educators using remote experiments was two-fold. First, one needs to teach (or reinforce from lecture) the chemical principles or theory components underlying the experiment. Second, one needs to make sure the student learns how to be comfortable with and effectively operate within that remote laboratory environment. With respect to the latter, we have found it is very important to match the experiment and student for the appropriate level of complexity. Computer manipulations and time invested must be reasonable for the eventual data obtained to seem worthwhile for the student.

As much as possible, the technology should not get in the way of learning and everything needed to carry out the experiment should be readily available for the student. That usually means a self-contained site that is seamless and intuitive from the students' perspective. The following are a few key features (summarized visually in Fig. 3) that have been incorporated into the CSRL environment.

1. Connection to the *instructor* for further address problems. Initially this was done asynchronously by email, but a chat feature on the website now provides more direct communication. This is a demand-side mediation approach designed to address specialized and real-time questions and needs of students. It is also intended to establish moral support for individual students through provision of a "teaching presence" connection [16].
2. Connection to *classmates* to encourage student-student interactions and to facilitate any group work.
3. Chemical principles *tutorials* to introduce students to the software, the instrument, and the particular experiment being undertaken. The purpose of the tutorial is to develop fundamental or remedial skills in the chemical principles, the instrumentation, and the online environment itself. General instrument tutorials consist of streaming videos that demonstrate use of instrumental software provided by the manufacturer to perform instrumental analysis. Interactive review exercises are also available to aid students in the recall of some of the key features of a particular aspect of the instrumental analysis software. Separate tutorials exist for particular experiments on a given instrument and may include downloadable text files.
4. *Qualifier exercises* to establish a minimum skill level for students before moving onto actual instrument access. This provides scaffolding for students to help them master content and skills. It also limits exposing expensive instrumentation to complete novices, which may have the potential for frustration on all sides.
5. *Scheduler* function to help students and instructors assign unique and secure instrument time to qualified operators. This is a logistical function designed to insure effective use of these resources.
6. Access to the analytical instrument to carry out *live experiments* on real samples submitted earlier. This places students in a discovery- or problem-based learning environment wherein they measure and collect real data. The intent is to reinforce upon students basic concepts and develop laboratory skills, while navigating potentially non-ideal results. Also, a web camera allows real-time viewing of the instrument during the experiment. The purpose is to make the instrument real for the remote student (i.e., seeing is believing).

7. *e-Logbooks* allow students to make comments and house data collected from the instrument. These "e-Logbooks" are typically submitted by students, either in whole or in part, for their laboratory report. The use of e-Logbooks allows for the facile handling of large data sets collected, so students are not immersed in the details of preparing tables by hand, but instead focused on learning the actual experiment itself. It also provides students with a vehicle to develop their note taking skills, which is an important part of scientific experimentation.
8. *Chemical databases* obtained mostly commercially from the instrument manufacturer and used as a reference library to compare with actual data obtained. This resource allows students to practice contrasting their measurements with more standardized or ideal results.
9. *Supplementary resource* materials provided for each experiment guide learners to use additional information or useful links for further study. This is a gateway into related online literature rather than a comprehensive source. The purpose is to use them as a resource for report preparation and to make students aware of the larger body of information available.
10. *FAQs* and Help sections address common problems encountered by students. This is a proactive supply-side mediation approach designed to efficiently answer commonly anticipated questions.

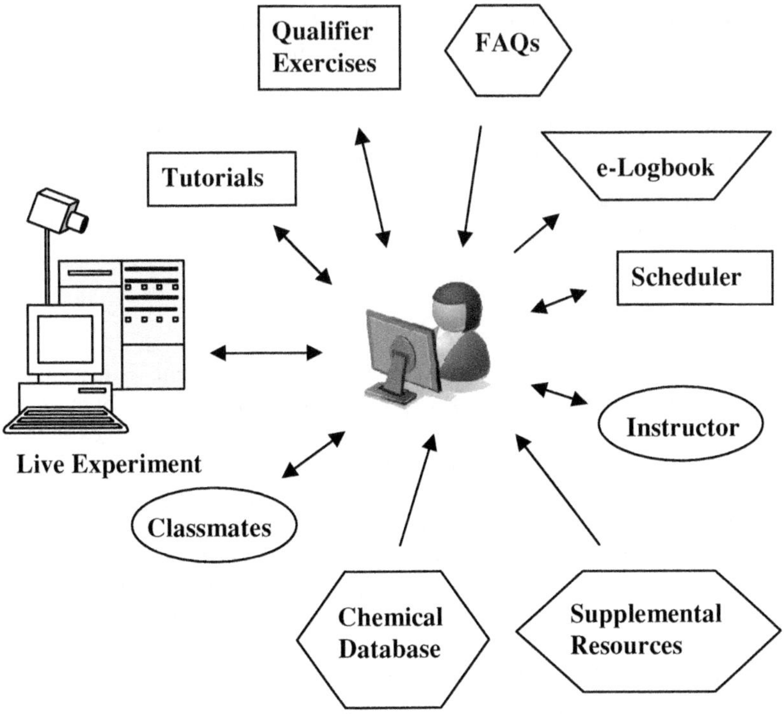

Fig. 3. Student-centred Features of the Chemistry Remote Laboratory

7 Student and Instructor Feedback

As we add experiments and incorporate changes we also formally and informally seek feedback from both students and instructors. The most recent feedback was a formal assessment by the capstone students in BTE 495 in their final report. Their recommendations for future development of the remote laboratory environment included an assignment drop box, online grade book, and a common work area feature to allow for groups to share and develop data and reports. These are excellent suggestions, which will no doubt be further pursued. At the moment, the NAIT–AU team has been adding individual features to the remote laboratory to create a high quality learning environment for the learner that goes beyond just access to the analytical instrument. With some hindsight, the obvious question is whether this environment might be better housed or developed within a learning management system (LMS). An LMS that is already being used by students at their institution for other courses would have a number of advantages including familiarity.

For completeness we include some highlights of feedback we have received that reflect our experience with using remote teaching laboratories in analytical chemistry.

1. Access and flexibility of the experiments is greatly appreciated by students. Also, the faculty likes making better use of existing instrumentation by employing them during non-business hours.
2. It is important to match the experiment and student for the appropriate level of experimental and remote environment complexity. (The number and complexity of computer manipulations must be reasonable for the eventual data obtained to seem worthwhile.)
3. The webcam is a must – seeing is believing. There seems to be a real need to view the instrument (even if it is just a box) to make it feel real.
4. Problems can be a good thing. Unexpected technical and experimental results in small doses can lead to valuable learning opportunities.

8 Conclusion

Our intent is to offer instruction and access to online experiments in a seamless package, one which is as self-contained as possible and can offer a supportive learning environment for the student. There are numerous practical advantages with flexible access for remote students or even sharing of instrumentation, but there are also educational advantages. Not only can it reinforce disciplinary course concepts, the experience itself is a learning opportunity for the student given that remote access is becoming more common for practicing scientists and engineers. The challenge for us (science educators familiar with our discipline, our students and the local environment) is to appropriately exploit ICT innovations. Remote teaching laboratories are being increasingly integrated into post-secondary science courses. We have explored some of the issues and considerations around doing this effectively and have offered our own approach with CRSL as an example. Remote teaching laboratories will continue to make significant contributions to the way we approach science education.

References

1. Cox, M.J., Baruch, J.E.F.: Robotic Telescopes: An Interactive Exhibit on the World Wide Web. In: Proceedings of the 2nd International Conference of the World Wide Web, Chicago (1994)
2. Penfield Jr., P., Larson, R.C.: Education via Advanced Technologies. IEEE Trans. Educ. 39, 436–443 (1996)
3. Kennepohl, D.: Remote Control Teaching Laboratories and Practicals. In: Kennepohl, D., Shaw, L. (eds.) Accessible Elements: Teaching Science Online and at a Distance, pp. 167–187. AU Press, Canada (2010)
4. Bailey, P.D., Garratt, J.: Chemical Education: Theory and Practice. Univ. Chem. Educ. 6, 39–57 (2002)
5. Ogot, M., Elliot, G., Glumac, N.: An Assessment of In-person and Remotely Operated Laboratories. J. Eng. Educ. 92, 57–64 (2003)
6. Doulgeri, Z., Matiakis, T.: A Web Telerobotic System to Teach Industrial Robot Path Planning and Control. IEEE Trans. Educ. 49, 263–270 (2006)
7. Fiore, L., Ratti, G.: Remote Laboratory and Animal Behaviour: An Interactive Open Field System. Comp. & Educ. 49, 1299–1307 (2007)
8. Cortera, J.E., Escheb, S.K., Mac, J., Nickerson, J.V.: Process and Learning Outcomes from Remotely-operated, Simulated, and Hands-on Student Laboratories. Comp. & Educ. 57(3), 2054–2067 (2011)
9. Lindsay, E.D., Good, C.: Effects of Laboratory Access Modes Upon Learning Outcomes. IEEE Trans. Educ. 48, 619–631 (2005)
10. Cagiltay, N.E., Aydin, E., Aydin, C.C., Kara, A., Alexandru, M.: Seven Principles of Instructional Content Design for a Remote Laboratory: A Case Study on ERRL. IEEE Trans. Educ. 54(2), 320–327 (2011)
11. Kennepohl, D., Baran, J., Connors, M., Quigley, K., Currie, R.: Remote Access to Instrumental Analysis for Distance Education in Science. IRRODL 6(3) (2005), `http://www.irrodl.org/content/v6.3/kennepohl.html`
12. Abdulwahed, M.: Towards Enhancing Laboratory Education by the Development and Evaluation of the Trilab: A Triple Access Mode (Virtual, Hands-on and Remote) Laboratory. Doctoral thesis Loughborough University, England (2010)
13. Orduña, P., García-Zubia, J., Irurzun, J., López-de-Ipiña, D., Rodriguez-Gil, L.: Enabling Mobile Access to Remote Laboratories. In: 2011 IEEE Global Engineering Education Conference (EDUCON) – Learning Environments and Ecosystems in Engineering Education, Amman, Jordan (2011)
14. Dziabenko, O., García-Zubia, J.: Remote Experiments and Online Games: How to Merge Them? In: 2011 IEEE Global Engineering Education Conference (EDUCON) – Learning Environments and Ecosystems in Engineering Education, Amman, Jordan (2011)
15. Connors, M.: A Decade of Success in Physics Distance Education at Athabasca University. Phys. Can. 60, 49–54 (2004)
16. Anderson, T., Rourke, L., Archer, W., Garrison, R.: Assessing Teaching Presence in Computer Conferencing Transcripts. J. Asynch. Learn. Net. 5(2) (2001), `http://www.aln.org/alnweb/journal/jaln-vol5issue2v2.htm`

Enterprise Portal Technology in Computer-Supported Collaborative Learning

Juha Ensio Kareinen and Jyri Pötry

Centre for Business and Engineering, North Karelia University of Applied Sciences,
Karjalankatu 3, 80200 Joensuu, Finland
{Juha.Kareinen,Jyri.Potry}@pkamk.fi

Abstract. Prevalent traditional e-learning platforms are monolithic systems, including most of the educational functionalities built in. In general, those solutions are often challenging and expensive to customize or develop for the customers' needs. Meanwhile, social web applications and collaborative tools have enhanced team working and online communication. These so-called web 2.0 features are available in enterprise portal software. This paper presents a case and a solution pilot where an open-source enterprise portal forms a platform for an e-learning environment. The aim was to test how social and other web applications may enhance computer-supported collaborative learning. The first results suggest that new web technologies allow new kinds of solutions and practices in e-learning.

Keywords: E-learning, Operations Management, Learning Management System, Enterprise Portal.

1 Introduction

In last decade, the web technologies and ways to use online services have changed significantly. The Internet has developed from static content views towards evermore interactive services, such as social networking sites, blogs, wikis, etc. This has brought us new prospects in online collaboration: it is possible to communicate, meet people, build networks and share information like it is possible in the physical world.

Today's web technologies seem to be mature enough to build flexible services for online learning, enabling collaborative working and providing versatile tools for the service users. However, even though the technology is available, the actual challenge is realizing a concept for e-learning as computer-supported collaborative learning (CSCL).

This paper will explore how these so-called web 2.0 features can be used for supporting collaborative learning. We are presenting an Operations Management themed learning environment pilot, based on enterprise portal technology and social web applications. The pilot was an outcome of R&D collaboration between North Karelia University of Applied Sciences and an e-learning service provider and developer, Arcusys Ltd.

Dieter Uckelmann et al. (Eds.): ImViReLL 2012, CCIS 282, pp. 54–61, 2012.

1.1 Computer-Supported Collaborative Learning and Web 2.0

At present, collaborative learning is one of the most popular learning types. Collaborative learning can be defined as a process where an individual is responsible for gaining knowledge through social interactions [9]. Computer-supported collaborative learning (CSCL) is an approach to learning where people use computers to interact with each other [10].

Nowadays, the Internet provides masses of social applications which can be used for knowledge sharing, interaction, collaboration and communications [7]. This evolved version of the Internet is called web 2.0 and it allows users to interact with information. Examples of web 2.0 applications include [6]:

- Social networking tools, such as Facebook and Linkedin, which allow people to build networks, share information and search user profiles
- Wikis, where multiple users can create and maintain content on a web site
- Blogs, where one keeps regular posting of ideas, opinions and news on a web site
- Twitter, the so-called micro-blogging site, where users send brief messages to others.
- Instant messaging, an application for real-time communication that connects users online with text-based communication.

These social networking tools cannot replace formal e-learning: structured learning is needed to deliver content, measure performance and track compliance [6]. Still, these applications can be used as basis for online communication to support formal e-learning.

1.2 Traditional E-learning with Stand-Alone Systems

Traditional e-learning is generally based on stand-alone system solutions, where all required e-learning functionalities are built into one application. "The traditional stand-alone learning management system (LMS) is built on an industrial age model. There are two specific problems with this model: first it is monolithic within a learning institution and second it is generic across learning institutions" [5]. These systems are plausible for e-learning when it requires hosting documents and text-based discussion boards. In more social e-learning they can cause restrictions to course realization or cannot provide all the features that the users would need [4, 5]. In addition, these systems are often provided by a single vendor. From their point of view it is not necessarily profitable to provide customized solutions. The resulted generic nature of the traditional LMS model means that the uniqueness of any given learning culture is undermined by a model that is more profitable for software developers. [3, 5]

2 Enterprise Portal Technology as a Basis for E-learning

Modern enterprise portal technology is versatile and wide spread with numerous use cases in managing accessibility, content and collaboration for different organizations. The majority of the enterprise portals are based on Java technology with standard APIs. Licensing models vary. From the point of view of public education, the platforms under open source – for instance, GPL or lGPL licensing – are especially interesting.

2.1 Enterprise Portals: Functions and Features

"An enterprise portal can be defined as a single point of access (SPOA) for the pooling, organizing, interacting and distributing of organizational knowledge" [1]. According to Raol, "the strength of corporate portals lies in ability to provide web-based access to enterprise information, applications and processes" [8]. This can make an enterprise portal a powerful corporate tool because any internal or external enterprise information can be accessed from anywhere with Internet access.

The basic function of an enterprise portal is, according to Raol, that "they leverage existing information systems, data stores, networks, workstations, servers and applications as well as other knowledge bases to give each employee in every corporate site immediate access to an invaluable set of corporate data anytime, anywhere" [8]. A role of an enterprise portal in corporate IT infrastructure is illustrated in Fig. 1.

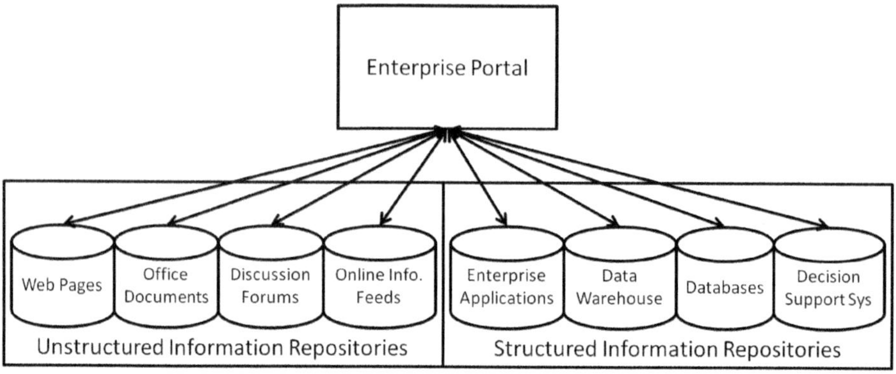

Fig. 1. Various information repositories that need to be integrated through the corporate portal [1]

However, managing this kind of architecture can be challenging. According to Aneja, easy expandability is essential for a portal system: "the portal framework needs to offer a plug-and-play capability that will allow additional functionality as the portal grows to meet future requirements [1]". In modern enterprise portals, plug-and-play capability is generally achieved with gadgets and portlets. "Gadgets are new application tools and services in the portal, provided via modular components. Gadgets provide the architectural construct to enable future extensibility without having to completely redevelop the portal" [1]. "Portlets are tiny applications that provide information and services from external systems" [2]. With portlets or gadgets there is no need to develop the source code of the portal when background systems change. Only the code of the portlet or the gadget has to be changed.

According to Raol, the one of the most common use cases for an enterprise portal is to act as integration platform: "common functions are the components that provide access to the range of disparate enterprise databases and information resources and the ease with users can set up personalized access to enterprise and external information resources" [8]. Raol also mentions the usage of portals as a platform for internal or external web content management or as a platform for collaborative working as common use cases: "in most enterprise portals, these functions may include, but are not limited to security, network, administrative tools, search, content management, collaboration, personalization, extensibility, easy to use and scalability" [8].

Portal technology can serve as a versatile platform for e-learning purposes. It provides the possibility to maintain web sites and content for formal e-learning courses but also allows integrating external applications and collaborative tools into one source. Integrated applications can be utilized in hands-on training and as collaborative tools diversifying and simplifying interaction between students and teachers.

2.2 Strengths and Weaknesses of Open-Source

An important benefit of open-source software is that there are often large and active developer communities. Sharing the results and new inventions is characteristic of open-source development. This allows software to evolve at the same rapid rate as the supporting technologies do. At best, the outcome is reduced costs in comparison to more a traditional system development. On the other hand, open-source applications tend to be released unfinished, which can result in higher costs than in commercially licensed software.

Open-source software can be very helpful for public organizations. In most cases, no commercial client applications or other software is needed when using services created with open-source software. Therefore, service customers are not bound to buy any commercial software which could cause restrictions for carrying out e-learning courses, for example.

3 Operations Management Pilot Learning Environment

As a test case, an e-learning environment for the operations management themed courses was built. The most critical requirements for the technology used were open-source licensing and the possibility to apply web 2.0 features. The first version was published in the autumn of 2011 and the concept was tested with a limited number of users. The finalization is under way and the first "official" courses supported by the pilot environment have begun in November 2011.

3.1 Requirements for E-learning Services

Technological requirements were to use open-source software in order to facilitate further development and to limit the licensing costs. The students and external customers had to be free to choose the software by which they use our services. Another technological requirement was to enable external web applications and the collaborative web 2.0 features. The aim was to contribute to team work utilizing and following some of the de-facto standards of the social Internet.

From the contents point of view, the idea was also to introduce the technologies to be understood in a learning environment. Operative IT systems such as enterprise resource planning systems (ERP) and product data management (PDM) systems are closely related to the field of operations management. The e-learning solution should thus offer easy access to the IT systems. The goal was to include as many practical elements as there are in traditional class room teaching.

3.2 Description of the Pilot Environment

Liferay was selected as the platform for learning environment. Liferay is an open-source enterprise portal system including the required features: the possibility to integrate external applications and built-in web 2.0 features.

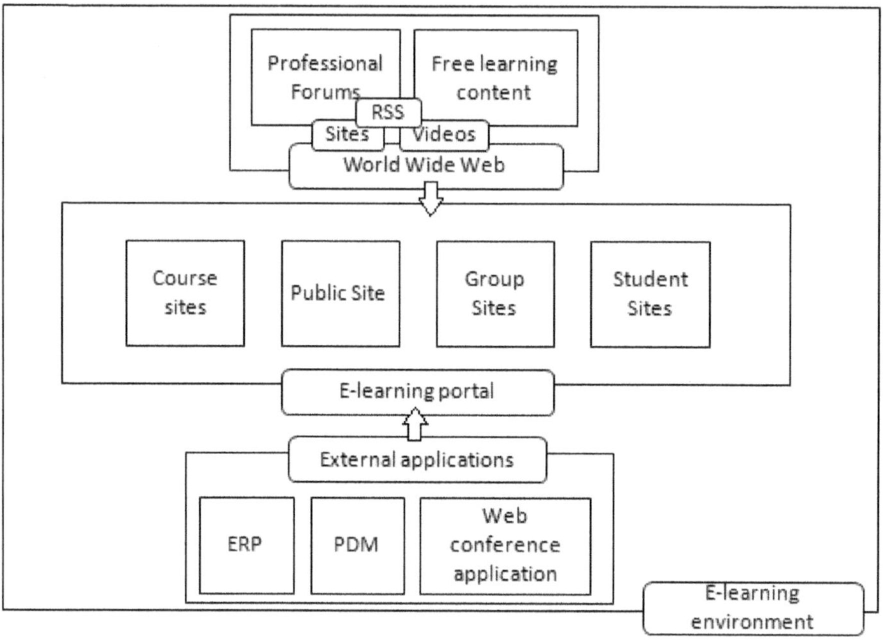

Fig. 2. Illustration of the e-learning environment. Functionalities and contents of the external applications are combined and presented on the e-learning portal site.

The pilot environment is illustrated in Fig. 2. It consists of a public site, course sites and different work sites of users, as illustrated in Figure 1. The public site is open to any Internet user. Registration for the courses, common information about them and coming events are found there. Each course has its own private site only visible to the registered users. Every student also has a site of his/her own where it is possible to maintain user profiles and to communicate with other users. Establishing communities (a type of group sites) for different tasks during courses is possible and allowed as well. The groups have their own collaborative tools, such as blogs, wikis, document banks, calendars, etc.

External contents, such as videos or animations, and applications, such as web-conferencing tools or ERP systems, are brought into the course sites with specific portlets.

3.3 Concept for the Pilot Course

A new user accesses the public site and registers for a course. In the registration, a user account and a student site are created for the user. Under the student site, the user can maintain his/her personal profile, keep a blog and a calendar, for example. The student site also includes a personal document bank where unreturned task documents or memos can be stored. Registration for a course also grants access to a specific course site.

The course site includes the formal e-learning structure and the learning contents. On the course site, teachers can maintain a calendar for course events and keep a blog to direct students. The users can access course topics, which are presented as text, illustrations, animations and videos. As Fig. 2 shows, the contents come both from internal and external sources. The origin does not affect the user experience. A message board and an instant messaging application are also available for discussion about course topics. After each lesson students are given tasks in different formats. The tasks can, for example, be returned into the course document bank or by answering questionnaires.

For group tasks, the students can form a group site, consisting of a web site and a document library. The group site creator may then invite other students to the community where they share the same resources. The group site can have the collaborative tools of its own such as wiki, blog and calendar that are only visible for the group members.

On the Operations Management course, a web client application of ERP and PDM systems are embedded into the course site. The students can access the systems through the portlets and fulfill required tasks.

The e-learning environment provides the students with numerous external sources of information. This includes various web content as well as professional contacts. Course tests are hosted with web form applications that are stored to portal database. Web forms are also used to gather feedback from course participants.

3.4 User Feedback

The Liferay e-learning pilot portal was tested with a small group of users. They were interviewed about the user experience concerning usability, features, importance of social applications and appearance of the e-learning portal. Most of the users had previous experience with some traditional e-learning software.

In general, Liferay portal was found easy to use. Creating new content and editing of pages were described simple. Majority of the users deemed features of Liferay portal versatile. Possibilities to communicate through different applications, to add different kinds of contents and to personalize the learning environment were mentioned useful. All of the users manifested that social applications can have an enhancing role in learning and education. The ideas were varying on which particular applications are the most useful, however. This cannot be regarded as limitation, because the personalization utility allows the users to apply the solutions according to their own preferences.

Criticism was expressed about some faulty configurations, a few complex or illogical functions and complexity of the user management. Also, some basic LMS functions were either still lacking or were regarded weaker than in the traditional solutions. In addition, some notices suggested that relatively much use training is needed for full utilization and user experience.

In general, the user feedback suggested that the appearance and the structure of the Liferay portal were clear when compared to other learning platforms. Most of the test users could see Liferay portal as an option for an e-learning platform and use it instead of their current LMS. Interactivity and centralization of information and applications were considered as important features of the learning portal solution.

4 Discussion

The enterprise portal technologies seem to be feasible in establishing a platform for e-learning purposes. Because of the possibility to integrate external applications into a single portal, it is relatively simple and fast to develop the learning environment. Different components can be added or removed without interfering or customizing the portal platform. New functionalities can be shared with the users via portal portlets.

Collaborative tools of the web 2.0 enable flexible team working. Service users can communicate as they have got used to in online services on the Internet. From a set of tools, they can choose how they work instead of following strict ways forced by the environment developers.

The idea of utilizing enterprise portals in combining learning communities, resources and services is not unique. There are already applications in the field and R&D efforts on better technical and pedagogical solutions apparent. In this paper, we have focused on the technical prospects (but do acknowledge that real benefits are gained via pedagogy). Open-source technology seems to be competitive in features when compared to commercial software. Benefits of using open-source software are open developer communities and that the users are not forced to a use of specific commercial software in their own devices.

New web technologies allow building e-learning courses that can be at least as versatile and practical as classroom lessons. In addition, flexible e-learning functions can always support and enhance traditional teaching. As a conclusion, there are several enabling, working and sparsely utilized technologies available. Their pedagogical advantages or disadvantages could be areas of further studies.

References

1. Aneja, A., Rowan, C., Brooksby, B.: Corporate Portal Framework for Transforming Content Chaos on Intranets. Intel Technol. J. 11, 21–28 (2000)
2. Boye, J.: Liferay Portal 4. EContent 29(8), 20–22 (2006)
3. Bush, M.D., Mott, J.D.: The transformation of learning with technology. Educ. Technol. Mag. 49(2), 3–20 (2009)
4. Coates, H., James, R., Baldwin, G.: A Critical examination of effects of learning management systems on university teaching and learning. Tert. Educ. Manag. 11, 19–36 (2005)
5. Culatta, R.: The Traditional LMS is Dead: Looking to a Modularized Future. Innov. Learn. (2011), http://www.innovativelearning.com/
6. Kane, K., Robinson-Combre, J., Berge, Z.L.: Tapping into social networking: Collaborating enhances both knowledge management and e-learning. VINE 40(1), 62–70 (2010)
7. Munguatosha, G.M., Muyinda, P.B., Lubega, J.T.: A social networked learning adoption model for higher education institutions in developing countries. On Horiz 19(4), 307–320 (2011)
8. Raol, J.M., Koong, K.S., Liu, L.C., Yu, C.S.: An identification and classification of enterprise portal functions and features. Ind. Manag. Data Syst. 102(7), 390–399 (2002)
9. Richards, A., Haythornthwaite, C.A.: The Sage handbook of e-learning research. SAGE Publications Ltd., London (2007)
10. Stahl, G., Koschmann, T., Suthers, D.: Computer-supported collaborative learning: An historical perspective. In: Sawyer, R.K. (ed.) Cambridge Handbook of the Learning Sciences, pp. 409–426. Cambridge University Press, Cambridge (2006)

Web-Assisted Laboratory for Control Education: Remote and Virtual Environments

Celina P. Leão, Filomena Soares, Helena Rodrigues, Eurico Seabra, José Machado, Pedro Farinha, and Sandra Costa

School of Engineering of University of Minho, Campus Azurém, 4700-058 Guimarães, Portugal
cpl@dps.uminho.pt, fsoares@dei.uminho.pt, helena@dsi.uminho.pt, {eseabra,jmachado}@dem.uminho.pt

Abstract. The Web Assisted Laboratory for Control (WALC) engineering on-line education is described. This on-line platform was developed by a multidisciplinary team. WALC enables the access to several virtual and remote laboratories experiences on automation, process control and numerical methods. The virtual environment enables the registered user to test different control systems (hydraulic, mechanical, electrical) working in open and close loop. This environment is simulated based on the numerical solutions of the differential equations that describe the real world systems. An industrial automation network is available allowing the user to test different programs running on PLCs. The remote access to the platform is confined to pre-registered users and under a booking system ensuring that only the registered user is on-line. Other functionalities are depicted. Based on user' feedback analysis, and since this platform was developed as a dynamic learning tool, their needs are fulfilled.

Keywords: Automation and Control Engineering, PLC Network, Remote Access, Virtual Labs.

1 Introduction

Due to strong demands on the practical training of engineers and, in particular, engineers in the field of control and automation, there are certain aspects that are fundamental, such as experimentation during the university education of engineers [1]. For this experimental education, physical labs with physical presence of students, have, traditionally, been used with the physical laboratory test benches where student's physical access was normal, current and each test bench was allocated to one single student, or a group of students, for a limited period of time.

It is obvious that experimentation *in situ*, with a plant or real object, cannot be replaced by a simulation or training simulators, due to the needs of education of engineers on this domain [2].

With the evolution of information and communication technologies it is possible that engineering students can access the laboratories and interact with the equipment

Dieter Uckelmann et al. (Eds.): ImViReLL 2012, CCIS 282, pp. 62–72, 2012.

at distance [3] and at any time via Internet or intranet. In this case we are dealing with remote access labs [4, 5].

Besides the use of remote access laboratories, it is also possible – due to the evolution of computing and the great possibilities that exist nowadays for the development of specific software tools – to create simulation environments that allow simulating behaviors of physical parts of automation systems [6, 7]. In this case we are dealing with virtual labs [8]. On engineering education virtual laboratories can, in a first step, replace remote access laboratories or traditional laboratories, but, in a second step of technical training, the real experiences must be performed and it is possible using remote access laboratories or traditional laboratories with physical presence of students, *in situ* [9].

Taking into account the aspects mentioned above, it is of crucial interest that institutions of higher education in engineering field – and more specifically in the domains of control and automation – incorporate in teaching activities. These subjects increase the use of virtual labs and remote access labs.

Another important aspect encouraging the development and sharing of virtual and remote labs is the resources sharing and optimization. As regards this aspect, we should consider the observations in the summary report of the NSF/CSS Workshop on New Directions in Control-Engineering Education [10], which are literally as follows:

"A shared laboratory can mean two or more departments sharing equipment and coordinating the development of experiments. It can mean the development of an integrated network of centralized laboratories, or it can mean sharing laboratories across campuses and across universities. Shared laboratories within individual colleges or universities, as well as shared laboratories among different universities, make more efficient use of resources, increase exposure of students to the multidisciplinary nature of control, and promote the interaction of faculty and students across disciplines. Shared laboratories also facilitate the horizontal and vertical integration of control systems concepts in the curriculum".

At School of Engineering of University of Minho, Portugal, four departments (Mechanical Engineering, Industrial Electronics, Information Systems, and Production and Systems) joined forces and resources for developing a platform that is used for teaching control and automation to their students. With this initiative, all the resources were grouped and optimized in order to allow that more students can access the same equipment, during 24 hours per day. This platform is being used on the context of virtual labs and also on the context of remote access labs.

Considering the context mentioned above, this paper describes the developed platform, by four departments of School of Engineering of University of Minho, highlighting the added value for all departments that collaborated on the project. The main focus of the paper is devoted to a small part of this big project concerning the exchanging of information between the user and the control system, taking into account security issues, for the controller and for the physical plant modules connected, in a closed loop, with the control platform. There are some rules and permission levels that must be defined, in order to assure the correct functioning of all system and this point will be discussed in detail in this paper. Also, the perception of the use of the platform by students is presented. This perception is the result of a questionnaire filled by students when they have acceded to the platform.

In order to achieve the goals proposed, concerning this paper, section 2 presents the developed platform, where the control and automation laboratories are briefly described, in section 3 the students' reaction towards the use of this platform is discussed and, finally, the conclusions are drawn in section 4.

2 WALC Platform Outline

As previously mentioned a multidisciplinary team of teachers and researches joined efforts and developed and made available a learning tool for a self-study, helping students to be successful on control and automation subjects. Thus, the Web Assisted Laboratory for Control Engineering on-line Education, WALC platform, emerges, facilitating the remote access, monitoring and control, to several laboratories, where students can complement their study by testing different case studies (Fig. 1) [11, 12]. The WALC platform can be access through the site walc.netai.net. The user (student) can discover theoretical and practical problems regarding automation, numerical methods and control subject as well as the relationship they may have in common. Social networks, such as Facebook and Twitter, can also be used by clicking on the respective available icon, to facilitate, to keep up to date and to share ideas among the users population.

Fig. 1. WALC main page (walc.netai.net)

2.1 Control Laboratory

In control teaching/learning process, WALC allows the student to choose the practical engineering problem and the control type to be implemented. Since these problems are simulated by a set of differential equations, the user is also able to test different numerical methods to solve the referred problems set together two areas of knowledge, process control and numerical methods.

The available real-world problems include: (1) DC Motor Control, (2) Temperature Control of a Classroom, (3) BakSim – Baker's yeast fermentation simulation, (4) LabSim – Virtual Laboratory for control problems, (4) Bar Control. The problems (1) and (2) are remote experiences and the remaining three are virtual experiences (Fig. 2). Notice that, all these available experiences were developed by undergraduate students as their final project work to get the engineering degree. The virtual experiences are always available and can run simultaneously. In opposition, the remote experiences only run with one registered user.

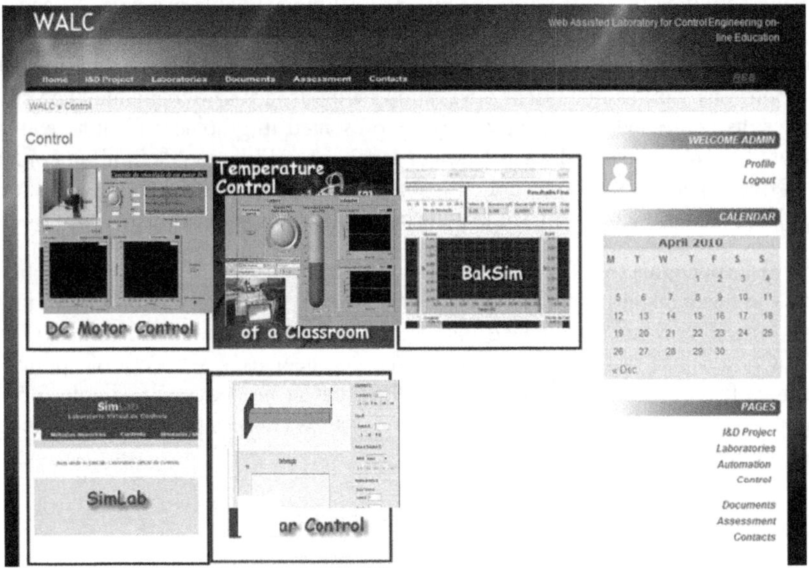

Fig. 2. Available control real-world problems in WALC

Remote Experience (1). The DC Motor Control is a remote accessed real-world experience, where the user can test a digital controller algorithm in motor velocity. The interface was developed in LabView from National Instruments. The experiment is controlled using a PID (Proportional, Integral and Derivative) algorithm. The remote user can test PID digital algorithms and parameters, change reference velocity values and register the motor output velocity profile [13]. The monitoring is performed by the graphical profile of motor velocity and the visualization through a webcam.

Remote Experience (2). The temperature control of a classroom system is capable of remotely monitoring and control the temperature of a physical space. The interface was developed in LabView custom application with a methodology of on-off control. The user can configure the system, set-point temperature and get overall monitoring via a touch screen display. The visual access to the experience is performed with a webcam and for the remote access it can be used any device supporting LabView environment [14].

Virtual Experience (3). The Baker's yeast Simulation, BakSim, is a virtual experience developed in LabView for virtual control and simulation of a bioreactor for baker's yeast fermentation, running in open or closed loop mode. It shows the most important system variables: biomass, ethanol, oxygen, glucose and dioxide of carbon. The BakSIM application allows the study of the fermentation process, the application of control laws and the effectiveness of several numerical methods [15].

Virtual Experience (4). LabSim is a virtual laboratory developed for numerical methods and process control. It is a user-centered and dynamic Web site to support the learning activities for university students. For all the available examples, process and control parameters (hysteresis or PID gains) can be changed during simulation test. Specific helping routines with theoretical support on the subject being studied, regarding control as well as numerical integration methods, are also available. The simulation results (state and control variable) are presented in graphical form as well as in animated form (animated draw of a water level control in a tank). SimLab contains a set of questions, randomly generated from a data base, permitting the user to test his knowledge and to identify difficulties during his learning process [16].

Virtual Experience (5). The problem defined as bar control, is a practical problem to study the mechanical properties of materials, namely the deformation measuring at a point on the surface of a flushed bar when subject to a controlled force [17]. The structures are part of our daily lives and the engineer uses a variety of applications that goes from bridges to skyscrapers. A flush bar at one edge is a simple form and example of a structure that can be used to define various parameters of its behavior when subjected to applied forces [17].

2.2 Automation Laboratory

The automation laboratory is composed by an industrial controller' network where different prototypes and scalable models, such as car parks, 'home automation' kits, or greenhouses, can be configured and tested by local or remote communication.

A locker or electrical switchboard was designed to be used as the core of the network process control: the Automation Network Locker (ANL) (Fig. 3). This locker contains a set of Programmable Logic Controllers (PLCs) that can be accessed through various communications tools: RS232, RS485, Mechatrolink, Profibus, DeviceNet and Ethernet [18, 19]. The access to the PLC's programming and network control is performed via Ethernet communication [20, 21].

Fig. 3. Automation Network Locker (ANL)

Fig. 4 illustrates the available ANL network where different communication protocols, signaled by different colors and numbers, can be implemented. Line (1) represents the DeviceNet communication and corresponding functions: exchange of data between control devices (PLC), pro-visions of security devices and control inputs/outputs. Line (2) represents the RS485 communication, containing two PLCs with different sensors. Line (3) refers to Mechatrolink communication to control the servomotor widely used for the transmission of power units. The Profibus, represented by line (4), links the inverter and a control device's inputs/outputs. And, finally, the Ethernet network, illustrated by line (5), contains the ANL network attached to two PLCs, a control device's inputs/outputs and a vision system. The Ethernet network allows the user to remotely access the system [22].

Students are able to write PLC programs using the CX-programmer (www.omron.com) on their own computer. The software CX-programmer is responsible for making the final deployment of PLC programs into the host-PLC through the Internet. However, CX-programmer does not provide any mean for integrating with WALC Laboratory Web site, through which students make the booking of the experiences. To overcome this problem, we have proposed a computing system architecture that performs the integration between WALC Laboratory Web site and the PLC

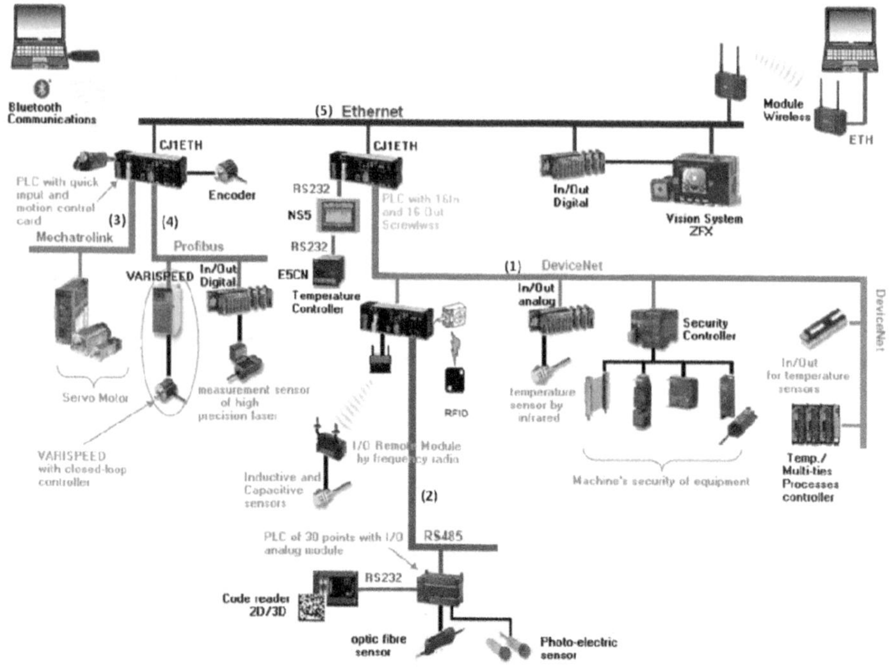

Fig. 4. ANL Network

closet: students develop and deploy their PLC programs using the CX-programmer component. The communication between the CX-Programmer component and the PLC closet is intermediated by a software component running on a local computer, which periodically configures the router for allowing or disallowing accesses to the PLC closet. This component consults the booking database in the lab Web site. If a booking is reserved for that student (IP-based authentication), the communication is redirected to the PLC closet.

The final version of the system uses a simple architecture, where both the laboratory's computer and the PLC closet are behind a router running the dd-WRT Linux distribution (Fig 5.).

Although not specified here, the PLCs are directly connected to the computer via a RS-232 cable. This can be used as a management or troubleshooting connection as it works with no or little configuration. Also, the PLCs are connected via Ethernet to the router and can be accessed from the computer via a private IP.

In this scenario, the router is the only device with a public IP address. It is responsible for allowing or denying TCP/IP connections to the PLCs. As already said, the local computer runs a software component that periodically queries the booking system database and uploads the forwarding configurations into the router. The database keeps the record of every registered student, with the IP addresses and the time intervals they are allowed to access the PLCs. As a rule, no user is allowed to access the PLCs for more than one hour per day. However, the administrator may override this rule when special cases are allowed.

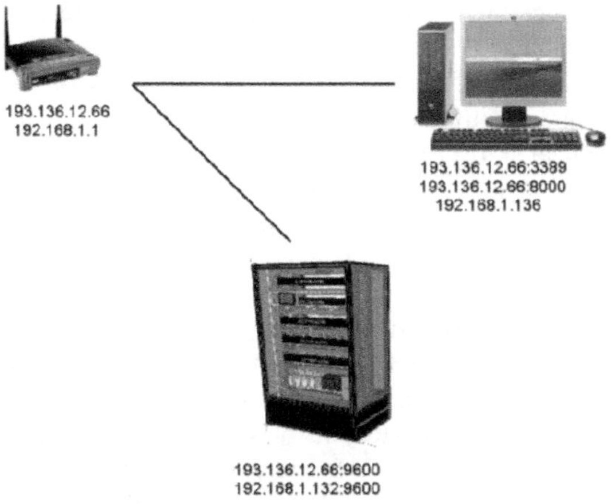

Fig. 5. Architecture to remotely access the ANL

The queries to the database and the router's configurations are managed by the system scheduler program, which triggers small batch files that are used together with the Tera Term Telnet client to configure the router on-the-fly. This configuration is done via a telnet or ssh connection to the router, thanks to the flexibility of the router's operating system.

Also, there are two permanent port forwardings to the local computer. Firstly, the TCP/IP port 3389 allows the local computer to be remotely administered. Secondly, as the local computer also serves the booking system, the port 8000 is always forwarding requests to the port 80 to the computer. However, in future work, the access to the booking system is expected to be available through a sub-domain of the Minho University, with no non-standard ports.

Most of the components used in the system are available as free and open-source licenses or are using open standards. The choice for open standards allowed us to develop an integrated architecture with almost no special technical skills on a specific tool or programming language. The final integration satisfied most of the requirements.

Having in consideration the previously developed work, we have used the following technologies: Apache Web Server, MySQL RDBMS, PHP, Wordpress CMS with a modified version of Ajax Event Calendar, Splinterware System Scheduler (Freeware), Tera Term SSH and Telnet client.

3 Insights on Students' Attitude

In order to understand how students react to this educational platform and how they see the effect of virtual and remote experiences on control and automation subjects as well as corresponding theoretical contents on their learning process, a questionnaire was developed and given to them at the end of each semester. The students were from Digital Control subject on the 3rd year of the Integrated Master of Industrial Electronics and Computers Engineering and from the Process Control subject on the 2nd Industrial Informatics undergraduate course [23, 24]. The questionnaire developed has six main parts: (1) characterization of the student, (2) identification of the work environment (operating system and identification of browsers), (3) educational kits: motivation (4), teaching kits: technical skills, (5) educational kits, and (6) transverse habits of students and characterization of the attitude during the project. To answer this questionnaire, each student needed only about 5 to 10 minutes. Data were collected on-line by computer through the platform, saved in a database and then analyzed [11, 23].

The first results help to state that the learning by experience engages components from the doing and the thinking. It was interesting to see that students, with the use of WALC platform, could understand and establish the relationship between the theoretical and practical concerns of the automation and control subjects.

The enthusiasm shown by the students on using the WALC platform motivates the project expansion of being hailed as a future work to include more remote laboratories in the platform and on the recognition of the multidisciplinary of this area of study.

4 Conclusions

The main purpose of this paper was the presentation and description of the WALC on-line platform. WALC is the Web Assisted Laboratory for Control (WALC) engineering on-line education and enables the access to several virtual and remote laboratories experiences on automation and process control education.

The virtual real-world experiences are always available enabling the user to test different control systems (hydraulic, mechanical and electrical), working in open and close loop. Another goal of WALC is to motivate students to understand how numerical methods tools can be applied in control studies. For each of the control experiences, the set of differentials equations that models the virtual experiences, are defined and the user can test the effectiveness of several numerical methods to obtain the final solutions, building synergies between the different acquired academic competencies.

The remote access to the available remote control experience is confined to preregistered users and under a booking system, ensuring that only the registered user is on-line. These experiences allow the user remotely monitoring and control physical parameters and reference values and register their influence on the output profiles.

For the automation laboratory, an industrial controller network for monitoring and access, was built based on Host Link protocol and Factory Interface Network Service protocol. The Automation Network Locker (ANL) designed and built, tries to reproduce the industrial network environment. The interface allows monitoring, reading, writing and changing parameters as well as memory locations, checking for errors and operation modes. The architecture used to remotely access the ANL is simple and most of the components used are available as free and open-source licenses.

Knowing that this platform was developed as a dynamic learning tool, the students' opinion becomes very important. Based on users' feedback analysis, it was possible to realize that, with the use of WALC platform, the students/users are able to establish a relationship between theoretical and practical concerns on both automation and control subjects, being the first contact with the industrial reality.

This work followed the improvements of remote laboratories as a supplement to traditional ones. Remote and virtual laboratories can be more rewarding for student learning and simplify considerably the logistics of laboratory resources.

Acknowledgments. The authors are grateful to the Portuguese Foundation for Science and Tecnology (FCT), through the R&D project PTDC/ESC/68069/2006 and the Strategic Project PEst-C/EEI/UI0319/2011, for financial support.

References

1. Horacek, P.: Laboratory experiments for control theory courses: A survey. Annu. Rev. Control. 24, 151–162 (2000)
2. Cooper, D., Fina, D.: Training simulators enhance process control education. In: ACC, San Diego, pp. 997–1001 (1999)
3. Poindexter, S.E., Heck, B.S.: Using the web in your courses: What can you do? What should you do? IEEE Control. Syst. Mag. 19(1), 83–92 (1999)
4. Bhandari, A., Shor, M.H.: Access to an instrumental control laboratory experiment through the World Wide Web. In: ACC, Philadelphia, pp. 1319–1325 (1988)
5. Bohus, C., Crowl, L.A., Aktan, B., Shor, M.H.: Running control engineering experiments over the Internet. In: 13th IFAC World Congress, San Francisco, pp. 25–33 (1996)
6. John Hopkins UNI applets, `http://www.jhu.edu/virtlab/virtlab.html`
7. International Conference on Computer Systems and Technologies – CompSysTech 2008, `http://jersey.uoregon.edu`
8. Fakas, G., Nguyen, A., Gillet, D.: The Electronic Laboratory Journal: A Collaborative and Cooperative Learning Environment for Web-Based Experimentation. Comput. Support. Coop. Work. 14(3), 189–216 (2005)
9. Poulis, D., Pouliezos, A.: Computer assisted learning for automatic control. In: IFAC Symposium on Advances in Control Education, Istanbul, pp. 181–184 (1997)
10. Antsaklis, P., Basar, T., DeCarlo, R., Harris, N., Spong, M., Yurkovich, S.: Report on the NSF/CSS Workshop on new directions in control engineering education. IEEE Control. Syst. Mag. 19(5), 53–58 (1999)
11. Costa, S., Leão, C.P., Soares, F., Rodrigues, H., Machado, J.: WALC – Web Assisted Laboratory for Control Engineering On-Line Education. In: 9th Portuguese Conference on Automatic Control, Controlo 2010, Coimbra (2010)
12. Costa, S., Soares, F.O., Leão, C.P.: Mixing Personal Learning and Control Engineering Education Environments. In: PAEE 2010 – Project Approaches in Engineering Education, Barcelona (2010)
13. Silva, V., Carvalho, V., Vasconcelos, R.M., Soares, F.O.: Remote PID Control of a DC Motor. In: REV 2007, Porto (2007)
14. Silva, M., Carvalho, V.: Remote System of Temperature Monitoring and Control. Int. J. Online. Eng. 4(4), 60–65 (2008)
15. Oliveira, H.C., Leão, C.P., Soares, F.O.: BakSIM – An application for control, monitoring and simulation of baker's yeast fermentation process. In: Brito, C.R., Ciampi, M.M. (eds.) The World Congress on Computer Science, Engineering and Technology Education Proceedings, New Engineering to a New World, Itanhaém, pp. 152–156 (2006)
16. Lemos, F., Leão, C.P., Soares, F.O.: SimLab: A Virtual Lab for Control Problems. An undergraduate Experience. In: Proceedings of Virtual University, Bratislava (2008)

17. Teixeira, L., Costa, O., Pereira, V., Leão, C.P., Soares, F.O., Restivo, M.T., Chouzal, F., Mendes, J., Campos, J.C.: Laboratórios virtuais: duas aplicações no ensino de engenharia de engenharia. In: Proceedings VII Simpósio Internacional de Informática Educativa (SIIE 2005), Leiria (2005)
18. Considine, D.M., Considine, G.D.: Standard Handbook of Industrial Automation. Champman and Hall, New York (1986)
19. Lee, K.: Sensor Networking and Interface Standardization. In: IEEE Instrumentation and Measurement Technology Conference, Budapest (2001)
20. Felser, M.: Real-time Ethernet – industry prospective. Proceedings of the IEEE 93(6), 1118–1129 (2005)
21. Kizza, J.M.: Security Threats to Computer Networks, Computer Communications and Networks Part II. Springer, USA (2005)
22. Leão, C.P., Soares, F.O., Machado, J., Seabra, E., Rodrigues, H.: Design and Development of an Industrial Network Laboratory. Int. J. Emerg. Technol. Learn. 6(S1), 21–26 (2011)
23. Leão, C.P., Soares, F.O., Carvalho, V., Vasconcelos, R.M.: Automation and Control Engineering Laboratory: Students Perspectives. Lectures Notes in Engineering and Computer Science 2185(1), 2017–2021 (2010)
24. Costa, S., Soares, F., Leão, C.P.: Students' perception on using WALC platform for Automation and Process Control engineering studies. In: Proceedings of the 14th International Conference on Interactive Collaborative Learning – ICL, Piestany (2011)

Developing Future Logistics Applications with the Saxony-Anhalt Galileo Test Bed

Klaus Richter and Olaf Poenicke

Fraunhofer Institute for Factory Operation and Automation IFF,
Sandtorstraße 22, 39106 Magdeburg, Germany
{klaus.richter,olaf.poenicke}@iff.fraunhofer.de

Abstract. The paper is giving a brief overview on the Saxony-Anhalt Galileo Test Bed as an integrated environment of several real labs and test beds for research and development in the fields of logistics and transportation. As the Fraunhofer IFF is one partner of the test bed consortium, specific research topics in the application field of telematics and logistics are addressed as well. These developments are all closely connected to industrial application as the Saxony-Anhalt Galileo Test Bed offers the platform for testing the technical integration of, for instance, video- and radio-based identification and localization systems into logistics process environments.

Keywords: Saxony-Anhalt Galileo Test Bed, Telematics and Logistics, Applied Research, Radio-Based Identification, Radio-Based Localization, Video-Based Localization, Virtual Reality Simulation, Secure RFID Bulk Reading, Freight Capacity Monitoring, Interchangeable Containers.

1 Introducing the Saxony-Anhalt Galileo Test Bed

The challenge of sustainably securing mobility in transportation and logistics was and is the focus of the state of Saxony-Anhalt's transportation and development policy. In its implementation of this goal, the Saxony-Anhalt Ministry of State Development and Transportation began consolidating applied research in transportation and logistics in a state initiative a few years ago. The development of Saxony-Anhalt Galileo Transport initiative has top priority. The initiative is intended to support companies and research organizations in Saxony-Anhalt, which develop new technologies, products, applications and services for overland transportation and logistics, in order to create new jobs in this high-tech sector in the medium and long term [1].

For that reason, the Saxony-Anhalt Galileo Test Bed was established with support of research and collaborating partners. The main goals of the Test Bed are the

- support for product development and the
- integration of new applications in transportation and logistics systems.

The Galileo Test Bed is coordinated by the Otto von Guericke University Magdeburg. The partners in the development and operation of the test bed with its various

Dieter Uckelmann et al. (Eds.): ImViReLL 2012, CCIS 282, pp. 73–80, 2012.

facilities are the Fraunhofer Institute for Factory Operation and Automation IFF, Magdeburg, the Magdeburger Hafen GmbH, the ifak – Institut für Automation und Kommunikation e.V. Magdeburg and the Hallesche Verkehrs-AG (HAVAG). Of these partners, the Fraunhofer IFF focuses on the application field of *Telematics and Logistics*. Further application fields of the test bed are *Navigation and Transportation* as well as *Communication and Transportation*. These application fields are mainly managed by ifak and HAVAG. Nevertheless, all three application fields are closely interconnected to transfer technologies and knowledge along these integrated fields of transportation and logistics.

As part of the state initiative Applied Transportation Research/Saxony-Anhalt Galileo Transport, several facilities of the Saxony-Anhalt Galileo Test Bed for Localization, Navigation and Communication in Transportation and Logistics were established and opened in 2010. In Magdeburg's Port of Science, the Development Laboratory that is especially employed for indoor process environments opened in March 2010. The Galileo Test Bed's Logistics Platform, a productive test environment in Magdeburger Hafen GmbH's Hanse Terminal, was opened in June 2010. This part of the test bed is especially employing outdoor focused applications for the identification and localization of logistics objects.

Fig. 1. Impressions of the Development Lab and the Logistics Platform in the Hanse Terminal (Photos: Dirk Mahler, Viktoria Kühne / Fraunhofer IFF)

Demands are mounting for automatic identification and localization of logistical assets to document and control logistical operations. This is the point of departure for important research at the Saxony-Anhalt Galileo Test Bed. For example, combined use of radio- and image-based identification and localization technologies to automatically determine the status of logistical operations by motion and state analyses can be used to generate significant synergies for safety and security tasks in logistics. Agreed upon among the partners in the logistics value added chain, these modules define essential elements of smart standardized logistics zones.

With the opening of the facilities of Saxony-Anhalt's Galileo Test Bed for Localization, Navigation and Communication in Transportation and Logistics, state-of-the-art development and test environments are now available to integrate established and new localization and identification systems in logistics operations. Nearly three million Euros, financed from funds from the state of Saxony-Anhalt and the economic stimulus package II, have been invested in the Galileo Test Bed, coordinated by Otto von Guericke University Magdeburg.

2 Identification and Localization Applications for Logistics Nodes

The Fraunhofer IFF lead manages research specializing in *Telematics and Logistics* at the Galileo Test Bed. For that purpose, the facilities of the development lab and the logistics platform are equipped with several technologies for identification and localization of logistics objects – thus radio based as well as video based. To provide the capability of seamless object tracking, several systems for indoor and outdoor localization are also part of the test beds equipment.

The logistics platform is a productive test environment in Magdeburger Hafen GmbH's Hanse Terminal, which provides numerous modes of handling for the implementation of localization and identification of technologies in outdoor operations. The equipment at this trimodal terminal for inland waterways, railroads and roads is primarily used to handle and transport delivered cargo. Reach stackers, a gantry crane, crane trucks that load heavy cargo and forklifts are used for trimodal handling among inland vessels, trains and trucks. The companies located on the port's premises handle primarily containers and wind turbine components. This logistics platform is a prototype for logistics operations in nodal points of intermodal transport chains, such as port premises and freight terminals.

The goal is to use new telematic technologies to modify terminal operations for increased demands of secure handling in order to boost productivity and auditing compliance. By that, competitive edges can be created for logistics nodes employing these new technologies. For identification and localization of objects on the Hanse Terminals premises, local radio- and video-based systems are developed and tested besides established GNSS systems. These local systems employing, for instance transponder technology or video surveillance, provide locating accuracies superior to GNSS-based systems. In particular, higher clock rates can be achieved by the radio-based local positioning system compared to GNSS systems.

Such technologies enable application developers to effectively meet the challenges facing logistics, such as localizing freight with the accuracy of pallet storage positions or localizing individuals for maximum occupational safety. Furthermore, motion trajectories can be continuously recorded under the conditions of a productively operating container terminal.

3 Applications for Video-Based Analyses

More precise analysis of nodal points is a crucial element for the continuous traceability of freight throughout the transport chain. Anomalies can quickly arise wherever freight changes its mode of transport. Surveillance systems are increasingly being used, even at transfer and warehouse facilities, in order to produce consistently high transparency for a secure supply chain. Video monitoring systems have become standard equipment in internationally operating logistics hubs because of auditing compliance. The Fraunhofer IFF and its partners' approach to research and development is to make these infrastructures usable for object identification and tracking in the logistics control center, too.

Video analysis in the local environment of a logistics hub is evolving into a precise method to localize coded, logistical assets in real time in the centimeter range within a georeferenced environment. The impacts of non-cooperative outdoor measuring environments, for example greatly varying illumination or weather phenomena, must be incorporated in the analysis algorithms in order to make technology industrially usable outdoors. Since it furnishes an outdoor environment and a productive process environment, the Galileo Test Bed's logistics platform serves as an important test environment for the necessary research activities.

Virtual reality (VR) simulation systems with extensive material handling model libraries augmented by models of radio and imaging systems are an important tool for the evaluation of analysis algorithms. The VR model libraries are used to engineer video systems on the port's premises. The individual cameras' fields of view are simulated in a VR model beforehand and disrupted to optimize the camera arrays and lens selection [2].

Fig. 2. VR-Tool for planning and testing of camera views (Fraunhofer IFF)

The Federal Ministry of Education and Research is supporting this research work in the project "Virtual and Augmented Reality for Maximum Embedded System Safety, Security and Reliability ViERforES[1]", which is part of its initiative for Cutting Edge Research and Innovation. The goal is to create and use virtual, georeferenced scenes based on knowledge of a real or planned logistical scenario including the surrounding situation, which support the assessment of an analyzed video system relative to a non-cooperative measuring situation. Virtual reality furnishes a freely manipulable test environment for the evaluation of image processing algorithms.

The Hanse Terminal serves as a comparative environment to identify the problems of non-cooperative measuring environments for video analyses and for reference tests to evaluate the image analysis algorithms developed. The productive process environment supports iterative improvements of VR applications and image analysis. Thus, the VR environments established in ViERforES provide a virtual lab of the existing environments of the real lab in the logistics platform of the Galileo Test Bed.

[1] ViERforES is funded by BMBF; project reference number: 01IM10002A.

The interconnection between these two Lab levels will be further enhanced within ViERforES, as monitoring functionalities will be developed to store single logistics process states in a VR model. By that, for instance, the status and locations of containers stored at the Hanse Terminal can be recorded by sensor data and transferred into the VR model. This model in turn can be used to forecast functionalities for the terminal management or for predicting, for example, clouding of camera views or radio transmission fields.

4 Smart Standardized Logistics Zones

An essential part for seamless tracking of logistics objects and standardization of logistics processes along the transport chains can be seen in RFID technologies. In the development laboratory of the Galileo Test Bed several systems for automated RF identification are established. The core technology for bulk reading of RFID tagged objects is the RFID tunnel gate, employing the Fraunhofer IFF's patented technology of a mode stirred chamber for RFID applications [3]. This technology has been tested in several setups in the development lab and has been successfully transferred into logistics processes of industrial partners. This Fraunhofer technology is employed in several applications, for instance at DHL and FIEGE premises for fashion logistics.

Fig. 3. RFID tunnel gate developed by Fraunhofer IFF in the Galileo Test Bed in operation at FIEGE warehouse in Ibbenbüren (Photo: Dirk Mahler / Fraunhofer IFF)

Furthermore, the Galileo Test Bed offers the environment to adapt the principle of the scalable RFID tunnel gate to customer specific requirements – ranging from applications for single parcels containing RFID tagged products up to applications for the identification of complete truck loads.

A further adaption of this principle is the smart container, which will become an important part for the seamless transparency along logistics process in multi-modal transport chains. These containers, automatically identifying the loaded goods on RF basis and updating their inventory, are furthermore enhanced with additional sensor technology. For that purpose the Galileo Test Bed has several container types available (reaching from a ULD to a standard ISO container) to develop and test the integration of different sensor technologies into the containers.

An essential new development is an optical depth analysis sensor, which allows gathering data of the physical loading status of a container. The current state of development in the Galileo development laboratory and its containers enables the automated analysis of free freight capacity within an interchangeable container which is used on a 3.5 ton vehicle. With that vehicle a mobile test equipment is given to the Galileo Test Bed. It is used to test and demonstrate the integrated usage of telematic technologies from RF identification, vehicle localization and automated load analysis.

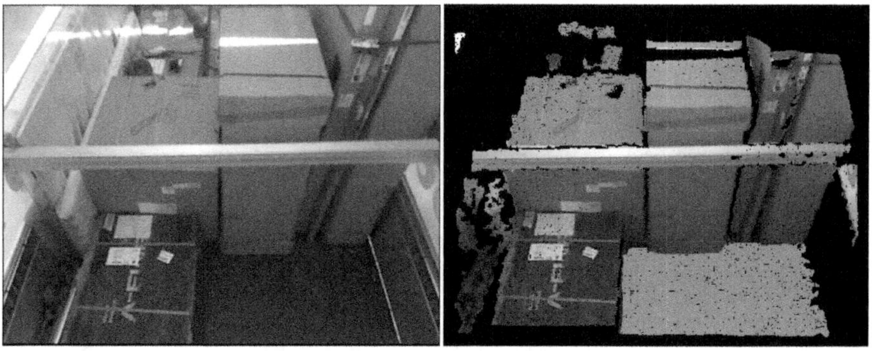

Fig. 4. Testing of automated detection of available freight capacities in a container using an optical depth analysis sensor (Photo: Fraunhofer IFF)

This information can, for instance, be used by transport providers to capitalize on free shipping capacities in already "rolling" transport ad hoc. Such application for load monitoring and localization as it is developed in the IMOTRIS[2] project, reports the current locations of vehicles as well as their available cargo capacity to a web platform, such as IMOTRIS. The system explicitly displays nearby requests for the transportation of shipments that fit in the free cargo capacities as potential ad hoc services. Service providers by that can boost their vehicle utilization and can pass a part of their cost benefits along to their customers [4]. First tests are carried out with industry partners within IMOTRIS, employing the smart container technologies developed and pre-tested in the Galileo Test Bed.

Such monitoring systems will help to organize logistics transportation – especially in inner cities – more efficiently in the future. The monitoring of the location and vehicle status is even of bigger importance for future use of electric cars and trucks in

[2] IMOTRIS is funded by BMWi; project reference number: 19G8021B.

terminating traffics to enable efficient routing and power consumption. In the next five years, the Galileo Test Bed will provide the development and test environment for a demonstration setup using an electric 3.5 ton truck with interchangeable containers that contain an additional battery pack and work as a range extender for the truck.

Fig. 5. Interchangeable containers for city logistics that are tested at Galileo Test Bed (Photo: Christian Richter / Fraunhofer IFF)

The demonstration setup will be built as part of the project MD-E4 (Magdeburg: EnergieEffiziente Stadt – Modellstadt für Erneuerbare Energien)[3] and tested together with industrial partners. The project will combine the new concept for electric transportation in city logistics with the use of sensor technologies for cargo monitoring that were developed in the Galileo Test Bed.

5 Conclusion and Outlook

Closely combining the development work in projects with the refinement of the Saxony-Anhalt Galileo Test Bed constitutes the basic strategy for the test bed and of all the research partners involved. In concert with the operators of international logistics hubs, a road map was developed when the Saxony-Anhalt Galileo Test Bed opened, which specifies future standards for telematics suppliers and integrators of logistics nodes, for instance, greater use of imaging technologies for the validation of radio applications.

In principle, the preparation, execution and evaluation of tests both under laboratory conditions in the Saxony-Anhalt Galileo Test Bed's development laboratory and under real conditions must be systematic, standardized and reproducible. Therefore, tests particularly focus on:

---

[3] MD-E4 is funded by BMBF; project reference number: 03SF0407D.

- the reliable operation of radio and image solutions under differing non-cooperative conditions, for instance movements and moving barriers,
- the verification of time and error characteristics, such as factors of uncertainty, quality of service and real time proximity,
- the coexistence of different radio and image applications, and
- the suitability of radio and image systems for functional applications with reliable IT for security and protection.

With its technical systems in its development laboratory and its partners' technical expertise, the Saxony-Anhalt Galileo Test Bed establishes outstanding conditions to develop and test reliable radio and image solutions as well as new concepts for material handling, such as the interchangeable container. The logistics platform in Magdeburg's Hanse Port, the development laboratory in Magdeburg's Port of Science and the other Galileo Test Bed facilities cover diverse aspects of holistic transportation and logistics operations to extensively develop and test solutions for clients [5].

References

1. Ministerium für Landesentwicklung und Verkehr des Landes Sachsen-Anhalt (ed): Galileo-Testfeld Sachsen-Anhalt – Landesinitiative Angewandte Verkehrsforschung (2009)
2. Richter, K., Poenicke, O., Nykolaychuk, M.: Galileo-Testfeld Sachsen-Anhalt – Funk- und bildbasierte Ortung im Hafenbereich. In: GeoForum MV 2011, Rostock (2011)
3. Richter, K., Kirch, M., Poenicke, O.: Sichere Pulkerfassung garantiert. Hebezeuge Fördermittel 51(5), 296–298 (2011)
4. Poenicke, O., Korzhynska, N.: IMOTRIS - Intermodal Transport Routing information system. In: 7th SoNorA University Think Tank Conference, Trieste (2011)
5. Schenk, M., Richter, K., Jumar, U., Herrmann, A.: Galileo-Testfeld Sachsen-Anhalt – Anwendungsschwerpunkte Verkehr und Logistik. In: Deutscher Logistik-Kongress. BVL, Berlin (2010)

Prototyping in Research Domains: A Prototype for Autonomous Production Logistics

Farideh Ganji, Marius Veigt, and Bernd Scholz-Reiter

BIBA, Bremer Institut für Produktion und Logistik GmbH at the University of Bremen, Hochschulring 20, 28359 Bremen, Germany
{gan,vei,bsr}@biba.uni-bremen.de

Abstract. The employment of simulation tools is the established method to analyze and validate research results. Generally, simulation tools can represent certain key characteristics or behaviors of a selected real-life or abstract system. However, these tools simplify approximations and assume relevant information. Today, most research areas are interdisciplinary. Therefore, the presentation and validation of collective results are a difficult challenge. Demonstrators and prototypes consider these difficulties and relevant factors, which can range from determining the required data sources to the employment of technological devices. The research domain "Autonomous Logistics" is such an interdisciplinary research domain, which focuses on the decentralization of decision-making processes. The showcase demonstrator "factory of autonomous products" is a development of this research domain. This paper introduces this showcase from the demonstrator perspective and proves the benefits.

Keywords: Simulation, Demonstrator, Prototype, Autonomous Logistics.

1 Introduction

In several research fields, simulation tools are suited to accurately analyze and validate the system performance. Simulation environments are important because they are a substitute for the "real world", thus allowing researchers to work in a safer and accessible environment [1]. Virtual simulations allow users to interact with the virtual world [2], which contains a platform of integrated software and hardware components. Multi-agent based simulation systems (MABS) [3, 4] are such a virtual simulation tool and provide a possible environment to prove the emergence of autonomous logistics systems. Primarily, the modeling of essential interaction between logistics objects is the crucial point in these systems.

Simulation systems that deal with specific parts of research areas often tend to leave out actual technological implementation and the relative necessary information source. However, the prototyping of hardware demonstrators enables the identification of adequate hardware, required interfaces and supplying of the needed system parameters.

In widely positioned research areas, the coherent presentation of the achieved results to the industry poses another difficult challenge. Demonstrators are a common

Dieter Uckelmann et al. (Eds.): ImViReLL 2012, CCIS 282, pp. 81–89, 2012.

way of solving this dilemma, because they combine several results in the same platform. The demonstrators are suited to visualize applicable results and potentially can be used as prototypes. The most important aspects of such a prototype are the cognition of the applicability as well as the limitation of the current technology. Furthermore, prototype systems allow further research demand in the same or other related areas.

The CRC 637 "Autonomous Logistics" currently investigates the paradigm shift from centralized control systems to decentralized logistics processes [5] and contains a wide area of research topics. We have developed several demonstrators, which have two specific tasks: the presentation of the research domain has to be comprehendible, and the feasibility of the resulted methods and concepts has to be verified in a real environment.

This article describes one of the demonstrators and investigates the benefits as well as the associated complexity considering real world aspects.

2 Autonomous Production Logistics

The increasing structural dynamic of production logistics enhances complex systems [6]. Different changes cause this trend, for example: short product life cycles, decreasing size of lots and the rising number of product variants [6]. This trend demands a new flexible control system in production logistics [7], which fulfills the specific requirements and maintains the efficiency of traditional large-scale production at the same time [8]. Therefore, it is advantageous to delay the differentiation of product variants [9]. In order to meet these requirements, the idea of autonomous cooperating logistics processes arose.

The main idea behind the autonomous cooperating logistics processes is to develop decentralized and heterarchical planning and controlling methods. In this approach, autonomic decision competencies are shifted to logistics objects, which plan and control themselves within the system. The previously passive work pieces become intelligent products, which have the ability to handle data, make decisions and communicate [10]. This is possible by applying software agents as digital representatives of the physical objects. They are able to interact with the production planning system, the processing stations and other intelligent products. Such complex networks of interacting intelligent objects are realized by multi agent systems [11]. This distributed decision-making allows to handle complexity and to react to environmental changes rapidly. The implementation of distributed decision-making requires the analyzing of the existing flexibility potentials in the field of production logistics.

2.1 Flexibility Potentials Enable Autonomous Production Control

Production logistics is categorized into the long-term production planning and the short-term production control [12]. Production planning contains the main strategic decisions, such as required capacities, assembly materials as well as an approximated schedule for following operations. The decisions during the production control include a precise definition of different final products and the order of their respective processing steps. In addition, the production control system arranges and operates the material transport. At the same time, the work-in-progress monitoring depends on the compiled production data and enables an intervention if required [12].

The analyzing of flexibility potentials in the manufacturing systems resulted in five essential types, namely allocation flexibility, machine flexibility, material handling flexibility, volume flexibility and operation flexibility [9]. These types are collected in the logical and the physical level of production systems. Autonomous products can use some of these flexibilities, which refer to the production control system. They can decide about the goal, the final product and the order of their production steps. The possibility to allocate customer orders depends on the processing status of the products and the customer orders. WINDT and JEKEN [9] visualize all alternatives in a product type corridor. Due to the processing status, the corridor narrows down. In the case of new emerging customer orders the corridor expands again (refer to Figure 1). Using the operation flexibility, the product can decide about the order of the production steps, which are left over for the appointed final product.

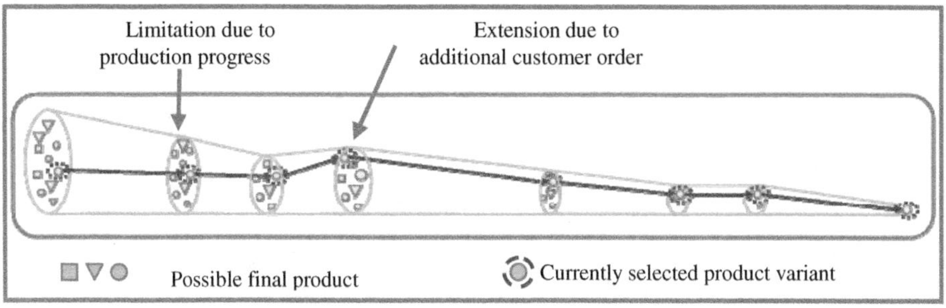

Fig. 1. Product type corridor (source: [9] p. 2)

2.2 Multi-agent Based "Factory of Autonomous Products"

There are already several results of using simulation environments to validate developed autonomous methods [13]. In the field of production logistics, we implemented a hardware demonstrator called "factory of autonomous products". The multi-agent based demonstrator covers the scientific works of the CRCR 637 as well as real-world components. The prototype system represents an autonomous assembly line for car tail lights. The final tail lights can be mounted with a colored, a black, or a transparent cover unit. The implemented methods allow the autonomous products along their production line to adapt the product variant as well as the next processing step to the dynamic changes of the boundary conditions.

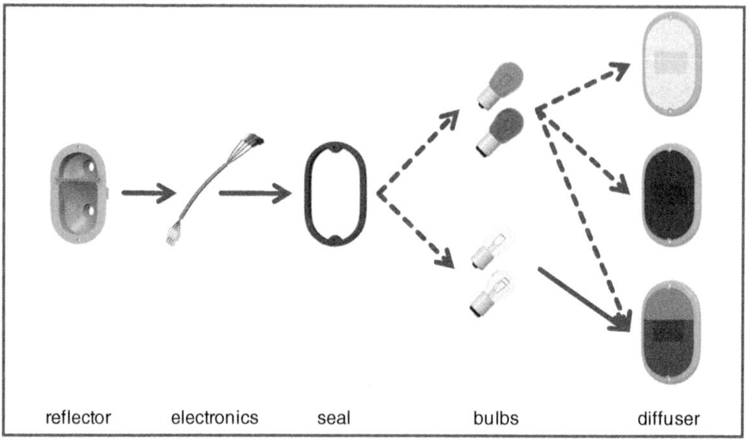

Fig. 2. Variant tree (source: [14])

The selected material flow system offers multiple transport routing paths. It is a monorail conveying system with self-propelling shuttles. The modular system design allows the product to change the planned route at any time [14].

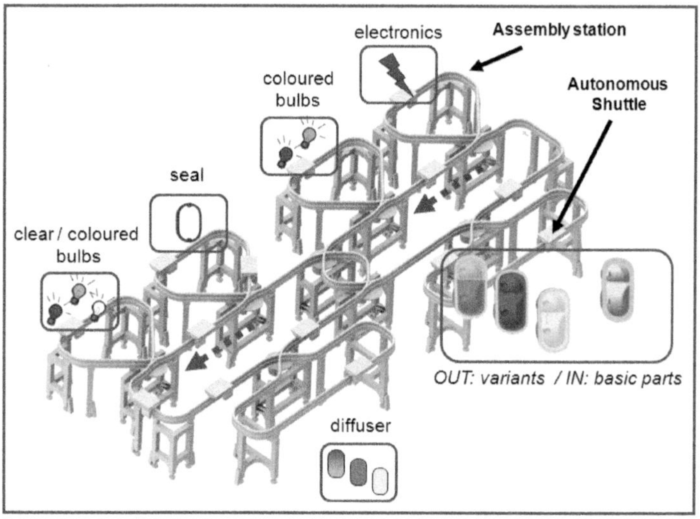

Fig. 3. The layout of the factory of autonomous products (source: [15])

To manage the complexity by working with the casted RFID tags, the used RFID technology was customized for our purposes and is characterized with a low frequency of 125 KHz. Since the oscillating circuits of RFID tags de-tune in metal environments, we used special antennas to compensate for the problem. Design, installation and testing of the hardware took about one year.

In order to explicitly identify the metal tail light units, RFID transponders are implemented during the casting process [15]. Associated with a software agent, the physical objects can manage their own production logistics. The synchronization of detection time by antennas and the control devices of the material flow system arrange the shifting of the work pieces to the desired destination. Furthermore, they can store and administrate the required data of processing. The casting pieces respond to disturbances and engine failure as well as short-term changes of customer requirements.

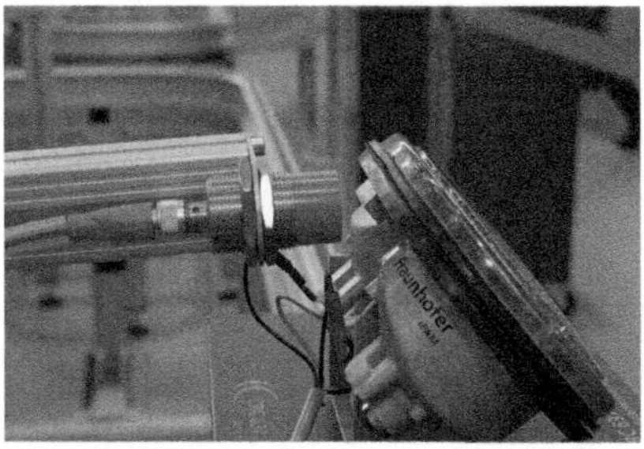

Fig. 4. Identifying of casting pieces along the production line via RFID

The scenario runs on an industrial material flow system, which has the required flexible construction technique. The self-propelled work piece shuttles have a carrying capacity of up to 12 kg. The design of the system allows a flexible routing change for the shuttles because of multi-directional movable switches. In this multi-agent system based implementation, the software agents represent all participating logistics objects. The so-called product agents occupy a central position in the scenario. The product agent employs the introduced method based on the autonomous product manufacturing cycle for assembly systems [9]. The currently possible variant and the scheduling to the next production step are determined by the implemented decision method.

Fig. 5. Moveable switches allow flexible routing change

The complete production planning is a central organized system. A flexible and rapid change in this system can be complicated, costly, and requires the flow back of updated system information, such as unexpected occurrences, in the system itself. The described demonstrator system implements the mentioned requirements. It can collect the relevant data, update and make them available.

3 Technological Requirements

Logistic objects require a useful combination of virtual and physical technologies to get the ability of autonomous production planning and to control themselves. In this chapter we summarize some of these technical requirements.

3.1 Identification of Logistics Objects

The tagging or embedding of auto-ID devices can guarantee the unique identification possibility of autonomous objects. In addition, the idea of autonomous control requires the management of environment information, which is relevant to control the processes. At run-time, the software agent saves required data. However, the life cycle of the product is definitely longer than the run-time of its responsible agent, which can cause a problem. A proposed solution that meets the combination of local data storage and unique identification is the RFID technology. The arrangement of antennas allows the localization of work pieces that are tagged with transponders. The low frequency range enforces a very short distance between the transponder and the antenna.

Due to different frequency ranges, the identification of capable technology depends considerably on the application area. Therefore, several connected application areas are constrained to apply the technology with the same frequency range. This fact points to an initial limitation of the feasibility within internal logistics. The system performance differs by reading and writing operations by different RFID-devices. Writing operations are most problematic. These factors complicate a modular implementing of interfaces to these devices. Consequently, the applicable hardware depends on the realized system. The verification of the pertinence is rather possible in prototypes as in simulation environments.

In summary, the unique identification of objects is the most important technology to implement the autonomous control. The RFID tagging has several advantages over other comparable technologies, such as barcode. The possibility of integration to the physical objects is representative of these advantages. In addition, the prospective data storage constitutes a further advantage of this technology.

3.2 Interaction between Logistics Objects

The continuous data exchange is an important requirement to implement autonomous control processes. The distributed decision-making process requires information on the state of the dynamic system data. This information refers to the number of customer orders as well as the local information handled by different objects. Unpredictable events are further dynamic changes. In addition to this information, the decision-making process also considers the static system data, such as configuration of product variants and the routing to the processing stations.

Multi-agent systems are suited to implement the described distributed decision-making and control system. The interaction between agents is organized via standardized agent communication languages and messaging systems [16]. Consequently, the modeling of the whole system at the multi-agent based simulation systems is a feasible approach. To begin, the modeling of interactions with the intended hardware is possible in simulation. However, due to different hardware performances, hardware testing and matching is crucial. In order to have a realistic validation of the research results, prototyping is the ideal solution.

3.3 Data Sources

The production planning system is the static data source for the production control. It contains the information about the configuration of final products, routing for material transport systems, as well as the number of customer orders. An interconnection between the production planning system and the distributed control processes is an essential requirement to implement autonomous logistic systems. The production planning as a management system can regenerate the configuration data by receiving the run-time data from the running processes. The previously management shifts to be a running process, which continuously revises the system configuration and provides it to other running processes. At the same time, every running process is converted to a data source for the management system as well as for all other processes. Consequently, an adequate interface to the current production planning and control systems is required. Prototypes implement such an interface, or they replace the management system and thus prototyping clarifies the question about individual data sources and the placement of needed arithmetic units. Simulation environments leave these questions open and position the data sources and the arithmetic units globally.

The described demonstrator replaces the production planning system with a configuration tool. This tool allows the configuration of several production scenarios and acts instead of production planning systems. Hence, the demonstrator presents a complete production planning and control system using the autonomous control methods. We implemented simulators for the applied hardware, which requires further development. Furthermore, a visualization tool allows an online Monitoring of the production processes. In the future, the tool can become a simulation tool for autonomous control systems in the field of production logistics.

4 Discussion

Simulation environments are best suited to prove individual results of research topics. The primary reason for prototyping in research domains is to build the right system by having an experiment platform. The importance of prototyping abounds when several research areas are associated. Research domains are interested in prototypes for demonstration proposes as well as using it as a simulation and experimental tool. However, the financial means for research domains are mostly limited. In addition, the hardware selection is associated with expenditure of time and tests. In order to accommodate for large-scale physical demonstrators, large spaces are required.

Demonstrators and prototypes are installed and maintained by staff. Unfortunately, after the expiration of the associated project, hardware demonstrators remain useless, unless there is a need for the prototype in the industry. Comprehensive demonstrators are an exception, due to the cost and the complexity.

The described demonstrator in this paper is a concept to realize autonomous control systems in the field of production logistics. The prototype emphasizes several of today's technical and methodical capabilities and initiates an approach to develop a simulation environment with a complete system performance. Such a simulation system provides a cost-efficient possibility to validate the autonomous control concepts in the future.

References

1. Sokolowski, J.A., Banks, C.M.: Principles of Modeling and Simulation: A Multidisciplinary Approach. John Wiley & Sons, Hoboken (2009)
2. Sherman, W.R., Craig, A.B.: Understanding Virtual Reality. Morgan Kaufmann, San Francisco (2003)
3. Davidsson, P.: Multi Agent Based Simulation: Beyond Social Simulation. In: Moss, S., Davidsson, P. (eds.) MABS 2000. LNCS (LNAI), vol. 1979, pp. 97–107. Springer, Heidelberg (2001)
4. Van Dyke Parunak, H., Savit, R., Riolo, R.L.: Agent-Based Modeling vs. Equation-Based Modeling: A Case Study and Users' Guide. In: Sichman, J.S., Conte, R., Gilbert, N. (eds.) MABS 1998. LNCS (LNAI), vol. 1534, pp. 10–25. Springer, Heidelberg (1998)
5. Hülsmann, M., Windt, K.: Understanding Autonomous Cooperation and Control in Logistics: The Impact of Autonomy on Management, Information, Communication and Material Flow. Springer, Berlin (2007)
6. Scholz-Reiter, B., Hildebrandt, T., Kolditz, J., Höhns, H.: Selbststeuerung in der Produktion – ein Modellierungskonzept. Ind. Manag. 21(4), 33–36 (2005)
7. Wiendahl, H.-P., Gers, D., Keunecke, L. (eds.): Variantenbeherrschung in der Montage: Konzept und Praxis der flexiblen Produktionsendstufe. Springer, Berlin (2004)
8. Scholz-Reiter, B., Windt, K., Kolditz, J., Böse, F., Hildebrandt, T., Philipp, T., Höhns, H.: New Concepts of Modelling and Evaluating Autonomous Logistics Processes. In: Chryssolouris, G., Mourtzis, D. (eds.) Manufacturing, Modelling, Management and Control 2004. IFAC Workshop Series, Elsevier Science, Amsterdam (2005)
9. Windt, K., Jeken, O.: Allocation Flexibility – A New Flexibility Type as an Enabler for Autonomous Control in Production Logistics. In: 42nd CIRP Conference on Manufacturing Systems, Grenoble (2009)
10. Meyer, G.G., Främling, K., Holmström, J.: Intelligent products: A survey. Comput. Ind. 60(3), 137–148 (2009)
11. Kirn, S., Herzog, O., Lockemann, P., Spaniol, O. (eds.): Multiagent Engineering: Theory and Applications in Enterprises. Springer, Berlin (2006)

12. Hackstein, R.: Produktionsplanung und -steuerung (PPS) – Ein Handbuch für die Betriebspraxis. VDI-Verlag, Düsseldorf (1989)
13. Scholz-Reiter, B., Görges, M., Philipp, T.: Autonomously controlled production systems – Influence of autonomous control level on logistic performance. CIRP Ann. Manuf. Technol. 58, 395–398 (2009)
14. Ganji, F., Morales Kluge, E., Scholz-Reiter, B.: Bringing Agents into Application: Intelligent Products in Autonomous Logistics. In: Artificial Intelligence and Logistics (AILog) – Workshop at ECAI 2010 (2010)
15. Pille, C.: Produktidentifikation, Intralogistik und Plagiatschutz – RFID-Integration in Gussbauteile. In: BDG-Fachtagung Gussteilkennzeichnung. Methoden und Datenmanagement – Praxisberichte, pp. V/1–V/4. VDG Akademie, Essen (2010)
16. Foundation for Intelligent Physical Agents (Fipa): Standard Status Specifications, `http://www.fipa.org/repository/standardspecs.html`

Agent-Based Emulation of an Electric Overhead Monorail System Pilot Plant

Testing and Virtual Extension of an Existing Test Facility

Willibald A. Günthner and Peter Tenerowicz-Wirth

Lehrstuhl für Fördertechnik Materialfluss Logistik, Technische Universität München, Boltzmannstr. 15, 85748 Garching b. München, Germany
{Guenthner,Tenerowicz}@fml.mw.tum.de

Abstract. Addressing the increasing need for flexibility in the in-house material flow concepts of decentralizing and modularizing material flow systems have become a popular research topic during the last decade. In this paper an agent-based approach for a distributed material flow control is described. The concept was validated on basis of an electric overhead monorail system pilot plant at the Institute for Materials Handling, Material Flow, Logistics (fml) at the Technische Universität München (TUM). An overview of the various modules and the control architecture of the pilot plant is given. The agent-based concept also offers new options for the simulation and emulation of material flow systems. These options are introduced in this article. Finally, an outlook is given on how the presented concept could support the collaboration between logistics labs.

Keywords: Multi-Agent System, Pilot Plant, Simulation, Emulation, Internet of Things, Distributed Material Flow Control, Intralogistics.

1 Introduction

Today's systems of internal material flow must cope with high demands: an increasing variety and shorter innovation and product life cycles lead to difficulties in predicting order loads and structures. The order-based production is becoming increasingly important. Against this background, material handling systems have to be dynamically adjustable and thereby cause the lowest possible cost. Approaches of modularization and decentralization both in the field of control software and in the design of conveyor systems promise great potential in terms of flexibility, robustness and reusability. Technological developments in the areas of sensors, data processing and storage, communication, locating and identification support this trend and enable new, innovative solutions for transportation tasks in logistic systems.

This paper focuses on a concept using a multi-agent system for controlling the modules of a material flow system represented by an electric overhead monorail system. Differently from current electric monorail systems, the trolleys as well as the track switches and cranes are autonomous elements featuring independent decision making and also flexible and extensible communication interfaces.

Dieter Uckelmann et al. (Eds.): ImViReLL 2012, CCIS 282, pp. 90–101, 2012.

1.1 Research at the Institute for Materials Handling, Material Flow, Logistics

The Institute for Materials Handling, Material Flow, Logistics (Lehrstuhl für Fördertechnik Materialfluss Logistik, fml) at the Technische Universität München (TUM) perceives itself as an open research institution aiming to significantly contribute towards the scientific progress in the areas of material flow technology and logistics. Insights gathered from fundamental research activities represent the basis for developing innovative solutions for current and practically relevant problems from research and industry. Along with aspects of technical logistics, the control and optimization of material flow processes by innovative identification technologies (RFID), the development of logistics planning by means of digital tools as well as the role of human beings in logistics are current research topics.

Embedded in this research profile, the elaboration of concepts for the decentralization and modularization of material flow systems marks one of the research focuses. The objective is to increase the changeability of such systems in its dimensions layout flexibility, flexibility concerning transported goods, throughput flexibility, and process flexibility [1]. Thus, in the project "MATVAR" (funded by the German Federal Ministry of Education and Research, BMBF) material handling systems for variable manufacturing segments in a dynamic manufacturing environment were developed [2]. Work on this topic was continued in the subproject "Changeable logistic structures in mini-factories", which was part of Collaborative Research Center SFB 582 "Market-driven production of customized products" [3]. In this context, a system for function-based modularization was developed which allows subdividing material flow systems in mechatronical units with a standardized functional and interface description. The developed concepts of modularization and distributed automation were successfully implemented in the institute's electric overhead monorail system (cf. Fig. 1) and subsequently further deepened – e.g., in the research project "Modular material flow systems for convertible factory structures" [4].

Fig. 1. Agent-controlled electric overhead monorail system pilot plant (*left*) with intelligent lifting units (*right*) at the Institute for Materials Handling, Material Flow, Logistics (fml) at the Technische Universität München (TUM)

1.2 Impact of the Internet of Things Paradigm on Intralogistics

The preliminary works mentioned above were integrated in another research project funded by the German Federal Ministry of Education and Research (BMBF) which addressed the impact of the Internet of Things paradigm on intralogistics [5]. In Co-operation with the Fraunhofer Institute for Material Flow and Logistics (IML) and corporate partners, a decentralized material flow control was implemented in several demonstration fields. During this project researchers at Institute for Materials Handling, Material Flow, Logistics (fml) developed methods for controlling and coordinating autonomous modules and transport units [1, 6]. Basic idea of the decentralized control concept is to assign the role of the material flow controlling instance to the transportable goods. To achieve their goals, the goods use transport services of the modules as well as supportive software services. The three basic units of a material flow system following the Internet of Things approach are thus

- **transport units**,
- **modules**, which provide (transport)functions, and
- **software services**, which support coordination between transport units and modules (e.g., directory facilitator) or system transparency (e.g., visualization).

For the communication between these basic units a multi-agent system was used. Depending on the number of potential recipients of a message, it will be exchanged peer-to-peer or via a blackboard system [7]. To each module, each transport unit and each service, a software agent is assigned, which supports a specific data exchange. Figure 2 shows how a decentralized control concept differs from conventional hierarchical material flow control.

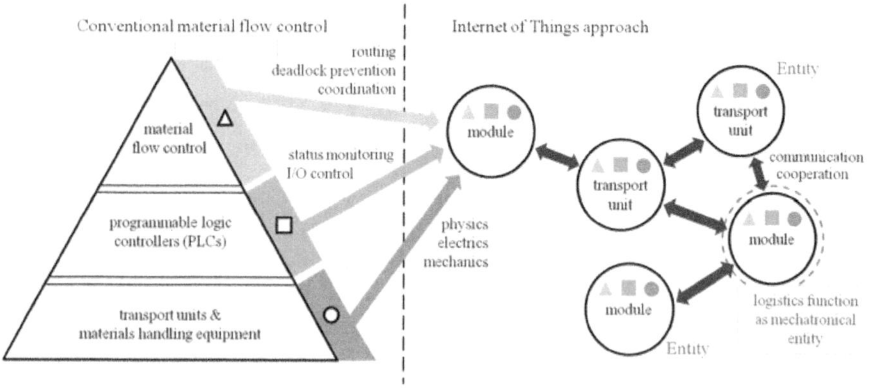

Fig. 2. Impact of the Internet of Things approach on intralogistics: autonomous, cooperating entities enable material flow systems without a strict control hierarchy

Functions that were hitherto located at different hierarchical levels of the control pyramid are distributed to modules and transport units. There are no higher instances of control like a material flow computer. This control principle has been implemented exemplarily in the above-mentioned electric overhead monorail system at the Institute for Materials Handling, Material Flow, Logistics (fml).

2 Electric Overhead Monorail System Pilot Plant

The electric overhead monorail system pilot plant (cf. Fig. 1) consists of a circular path with three switch points, a single-carrier crane and two single-carrier trolleys. The trolleys can automatically and manually supply and dispose goods to and from storage and conveying locations. Roll conveyors and a mobile rack are available to demonstrate the linking possibilities. The system can transport VDA small load containers of up to 50 kilogram. For handling VDA containers, an intelligent lifting unit, capable of compensating positioning inaccuracies of up to several centimeters, has been developed and patented.

2.1 Hardware Platform for the Modules

For the purposes of the Internet of Things approach, trolleys, crane and switches have to be seen as mechatronic modules, each equipped with an own computational unit. In contrast to the computational units of "software services" and "transport units", computer systems assigned to the entity "module" must be able to cover two areas of very different requirements:

1. Control and monitoring of physical processes in real-time
2. Communication/interaction with other entities and decentralized decision-making.

In the second case, no hard real-time capability is required, but a more complex logic. For this reason, different suitable programming languages were used. At plant level, programs were implemented according to IEC-61131-3, a standard for programmable logic controllers (PLCs). Only the guaranteed cycle time provided by a PLC makes it possible to control fast processes reliably. More complex tasks (e.g., agent communication) were programmed in PC-based environments and operating systems (Windows, Linux, etc.) with high-level programming languages like C# and Java. These provide useful mechanisms of object-oriented programming, such as encapsulation, inheritance, and polymorphism. In addition, many large libraries are available for various applications and thus reduce the programming effort. The major disadvantage of PC systems in terms of automation, however, is that they are usually not capable of real-time.

So called embedded PCs combine PC- and PLC-platforms and are both running a common operating system like Windows or Linux as well as a soft PLC that provides a real-time run-time environment for IEC-61131 applications. These combined software environments can communicate through various interfaces. Thus, embedded PCs enable the implementation of the two-layer architecture which is necessary for building the module agents. All higher tasks are assigned to the software agent that is created with high-level languages. The control and monitoring of inputs and outputs can be programmed in traditional PLC languages and outsourced to the real-time capable machine control level. The machine control is subordinated to the software agent and has no interfaces to other modules (cf. Fig. 3).

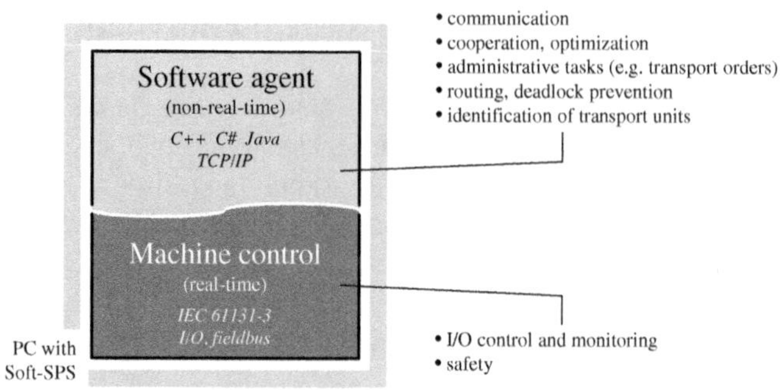

Fig. 3. Two-layer architecture for modules [1, p. 28]

2.2 Control Architecture of the Pilot Plant

In the test plant a JADE/LEAP software agent [8] and an IEC-61131-3 control program are assigned to each module. Both programs are executed on an embedded PC (cf. Fig. 4). In this concrete scenario the two trolleys and the crane have got their own computers while the three switches share a single hardware platform for reducing the cost of computing hardware. RFID readers for the identification of transport units were not installed in the electric monorail system. The mediating middleware is implemented as a collection of Java classes that are instantiated in each agent. It transfers information between the agent and the machine control or between the agent and an emulator according to the configuration stored in a XML-file (eXtensible Markup Language).

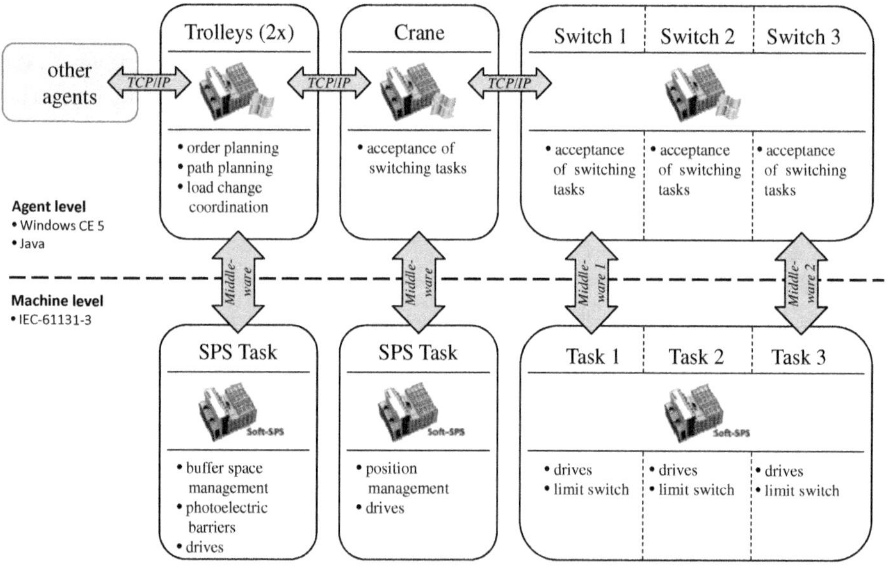

Fig. 4. Control architecture of the electric overhead monorail system pilot plant [1, p. 145]

2.3 Functionality and Processes

The trolleys periodically retrieve open transport orders from the blackboard (cf. Fig. 5). Given that a trolley gets a new executable job, it first checks if other free trolleys exist in the system and, therefore, a negotiation is necessary to allocate the job. This information is accessed through a directory facilitator (provided by the JADE framework).

If there is no other free trolley available, the trolley takes the order and enters a corresponding note on the blackboard. If a negotiation is necessary the involved trolleys in a first step calculate their bids which result from the length of the shortest path they can reserve for reaching the start point. Both trolleys write their own bid on the blackboard and retrieve all existing bids after a defined time of five seconds – thus, each vehicle gets an overall view of the auction and may recognize whether it has won the negotiation. Since the transport units in this scenario are not equipped with their own software agents, the order disposition must be controlled in the form of a non-moderated auction between the trolleys supported by the blackboard system. The winner reserves the order on the blackboard using its name (which prevents duplicate reservations) and then starts the order processing.

The calculation of a path from the current location of the trolley to the roll conveyor, of which the container is to be collected, is carried out with a Dijkstra's algorithm [9]. This algorithm takes the static system topology into account, which was retrieved by the agent during initialization. On the other hand it also considers the dynamic state of the system – such as the location of other vehicles, path reservations or manually entered locks. All this information is available on the blackboard, which guarantees the actuality and accuracy of the data. As soon as the trolley has found one or more pathways and opted for the shortest, it sets reservations for the route to be traveled. This is necessary to avoid collisions with other vehicles, especially from the point that the trolleys can move in both directions within the network. Once a route or a switch is left, the corresponding reservation is immediately deleted. During the drive, it can be necessary for the trolley to communicate peer-to-peer to a switch agent or the crane agent for realizing its planned route.

At the start point of the job, the trolley checks if load transfer coordination is necessary. If a roll conveyor for the corresponding coordinates is listed in the directory facilitator (DF) it is assured that both modules are in a safe condition for the load transfer by a simple exchange of messages between the modules' agents. The load-carrying unit of the trolley is descended and after gripping the container raised again. If the trolley is just equipped with a manual load-carrying unit, an operator must perform the gripping operation and then lifts the load. The successful completion of the load transfer is confirmed by a further exchange of messages between trolley and conveyor. Now, both of the modules can continue with their order processing.

The trolley now moves to the final destination. If the destination is blocked or reserved by the second trolley, the first trolley tries to give the other trolley an evasive command. When the route is clear, the process of path planning and reservation, transportation, and load transfer is repeated (cf. Fig. 5). After the mission is accomplished another message is send to the blackboard. Thus, an operator can get an up-to-date overview at any time using the visualization or the order management service. These services show – among other things like e.g., track reservations – which orders have been processed orders and which are still open.

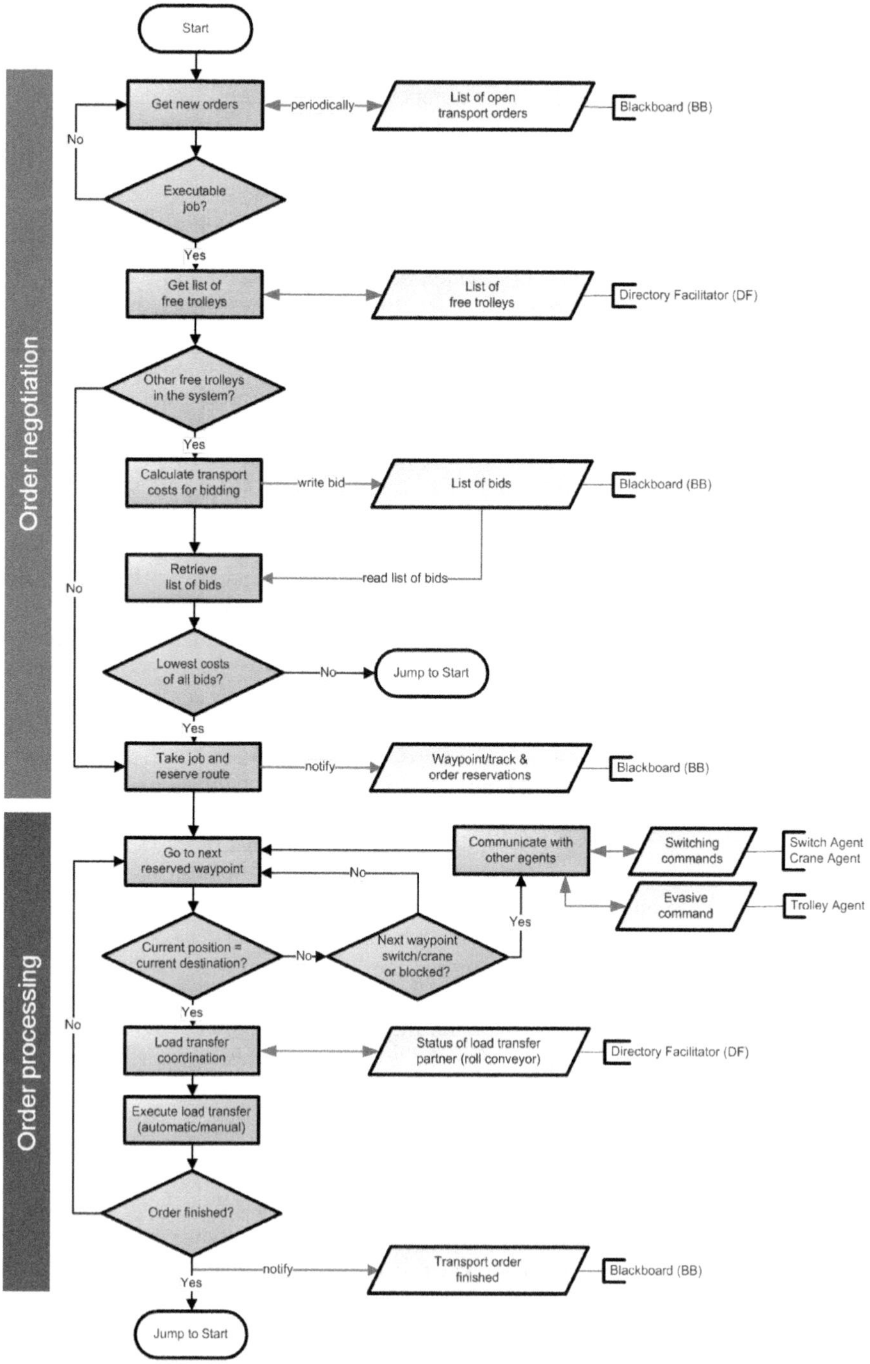

Fig. 5. Flow chart showing the general workflow of the trolley agent

2.4 Findings

From the software view, the implementation of a completely decentralized, agent-based control for an electric overhead monorail system demonstrates the simple interoperability and integration of programs in various run-time environments. In the described system, Java module agents on Windows CE computers communicate with net service agents (blackboard, visualization) running on a Windows XP-based PC and their associated IEC-61131-3-based machine control. The material flow control implemented some complex functions: order disposition, path planning and reservation, and the ability to automatically generate alternative (evasive) jobs and thus to ensure a deadlock-free operation. All these algorithms are independent of the topology, the order structure and the number of vehicles in the system. Thus, an extension or modification of the system is possible at any time. Faulty or blocked routes are recognized by the trolleys and automatically bypassed if alternative routes are available. This results in a robust and flexible system.

3 Simulation and Emulation of an Existing Pilot Plant

Not all aspects regarding the control of complex material flow systems could be investigated in the experimental hall of the Institute for Materials Handling, Material Flow, Logistics (fml). Hence, in addition to the execution of experiments on the pilot plant, a simulation model was used, which is particularly well suited for the analysis of highly dynamic processes in material flow systems. Because of the interaction between many different material flow components and the stochastic influences on the overall system, analytical tools are barely suitable. Hence, the simulation technique is often the only applicable method.

An important special application of the simulation is the so called emulation. When emulating, the machine control layer of the modules is replaced by a Java agent which imitates the functionality of the machine control (e.g., change of position). In this way, the software functionality of a module can be tested in a virtual environment without the mechanical or electrical components of the conveyor system.

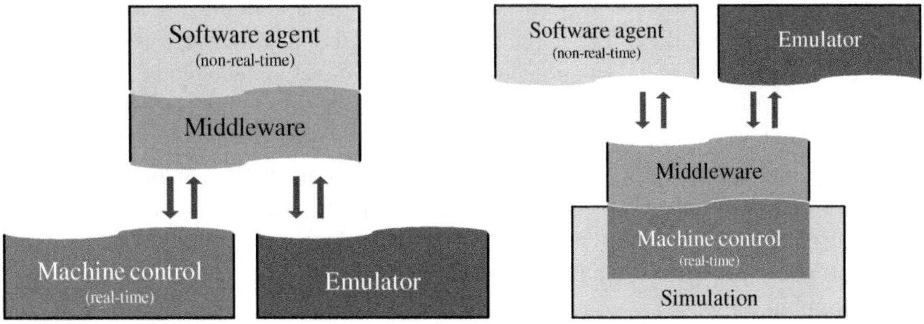

Fig. 6. Basic internal structure of an entity (*left*) and system design for the testing of machine-oriented software (*right*) [5, p. 190]

The coupling between software agent and emulator as well as between software agent and machine control was implemented using the abovementioned middleware (cf. Fig. 6, *left*). This middleware enables the replacement of the emulator by the control software of the real machine by formally describing the variables to be exchanged and mediating between the two software partners. The creation of an emulator for the machine control system in turn needs accurate data, which is derived from the interaction of the (real) machine control with the simulation environment (cf. Fig. 6, *right*).

3.1 Agent-Based Emulation

Regarding the use of simulation and emulation models, there are significant differences. Simulation models are used for extensive experimentation by many simulation runs, so that alternatives can be compared with each other by parameter variations. In emulation models, this is possible only in a limited way, because their execution speed is often restricted to real-time by the condition of the system to be controlled.

For testing the machine control, the software agent is replaced by a suitable emulator (cf. Fig. 6, *right*). This procedure results in entities whose entire software programs, including the machine control, have been extensively tested in advance. The emulation model should behave as far as possible analogous to the real system and must closely imitate the behavior of individual system components. If the representation of the material flow system is sufficiently accurate in the emulation model, the emulation provides a powerful tool which can be used at different stages in the life cycle (planning, implementation, operation) of a material flow system.

Emulation enables computer-based tests of the entire control software. Thus, the experiment implementation is easier and can be carried out faster compared to using the real system. It is also to consider that if changes, updates or extensions of the control software occur at a time when the real plant is already in operation, the emulation model can be used to test the functionality of the modified control software before using it in the real operation. On the other hand, effects of changes/extensions to the real plant can be analyzed by minor adjustments of the existing simulation model.

One of the main advantages of agent-based, decentralized control compared with conventional control architectures is that the control logic can be reused in form of an agent kit for other projects. In terms of simulation, this also means that the control system is already available in the planning phase – assumed that it was implemented in previous projects or in an project-independent development phase. It should be noted that software agents can communicate through simple interface configuration with both real and simulated logistic systems without additional programming effort. Therefore, the creation of an emulation model is technically possible at any time in the system life cycle. There are two options for the simulation of the system to be controlled, i.e. the behavior of the physical plant components: The use of a conventional (proprietary) material flow simulator and the use of an agent-based emulation kit.

Using an established material flow simulator requires a direct assignment via interfaces between the software elements of the simulation and the control agents who take all strategic decisions. It is important to ensure that these interfaces to the agent system can be disabled with minimal effort. In this way the software agents can be replaced by an also simulated control system. The physics of the simulated system, which are e.g., used in the planning stage for emulation, can later be reused during the implementation phase for a complete simulation of the overall system. The needed exchange of information can be implemented via sockets (e.g., using the TCP/IP protocol) or a (real-time) database with SQL interface. The advantage of this approach is the high degree of achievable accuracy for the simulation system mechanics.

When using an agent-based emulation kit, an emulator agent is assigned to each software agent. This emulator agent reflects the function of mechanics and machine control level (cf. Fig. 6, *left*). Since the control logic consists of a standard kit of software agents, a standard building block for emulators, which are assigned to the control agents in a direct 1:1 relationship, can be developed. These emulators can be executed within the same runtime environment as the control agents and thus require no external simulation tool. A disadvantage of this method lies in the effort to create the emulator kit. Functions like the calculation of logistical indicators, such as throughput, as well as mechanisms for the synchronization of the numerous emulators have to be implemented manually. However, for the data exchange between the control system and the emulator agents, no special communication blocks or additional interfaces are needed because all communication is ensured through the mechanisms of the employed agent system.

The advantages of using emulation for the testing of control systems can be summarized as follows:

- Support for more efficient debugging/troubleshooting without affecting the operation of the real plant
- Shorter launch time for control systems
- Avoidance of effort and errors due to repeated implementation of algorithms in the simulation model during the planning phase and in the control system itself
- Usage as a training tool for learning about the system control
- Support for creating functional requirements and specifications for the control system
- Support for creating control strategies of the control software
- Analysis of impact of changes in the control software or in the real plant
- Possibility of training self-learning software agents and optimizing them regarding their usage in actual operation.

3.2 Mixed Operation of Real and Simulated Components

The communication and control architecture further allows for the integration of virtual trolleys and tracks into the system. In this way, a mixed operation of real and simulated modules is made possible, thus allowing a better evaluation of planned alterations or extensions to the material flow system. In this case, the mechanics of the new sectors are depicted in a simulation environment, while their control is already implemented with the software agents, which will later be used in the operating phase. In this scenario, these new software agents are executed within the same run-time environment as the already instantiated agents (which control the real system) and are able to interact with them. This represents a mixed operation of real and emulated components which can be seen as a preliminary testing of a planned system expansion (cf. Fig. 7). Thus, important information about the future performance of the extended system can be provided without requiring a simulation or emulation of the already existing sectors of the plant.

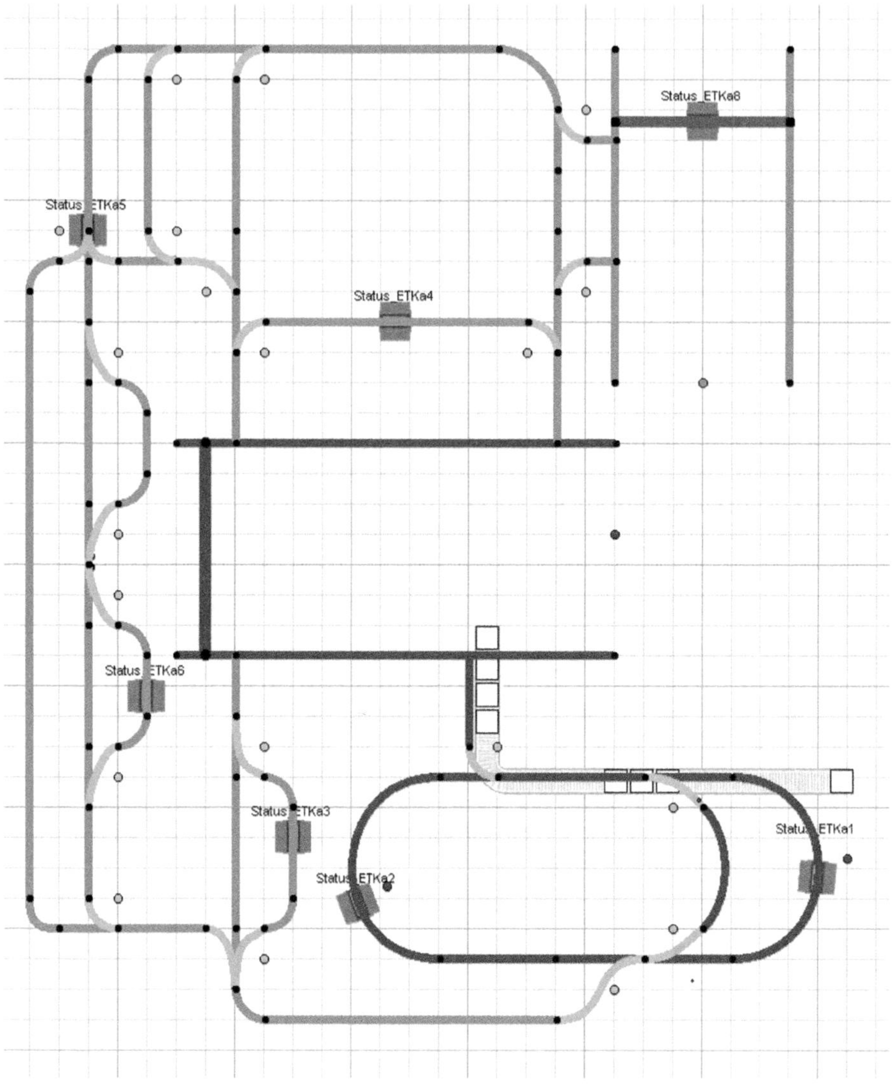

Fig. 7. Mixed operation of a real electric overhead monorail system (*dark gray*) and a simulated, virtual extension of the plant (*bright gray*) during the operating phase. The rule-based visualization shows plant layout as well as trolley positions and states.

4 Outlook

Agent-based control and emulation of pilot plants which was illustrated in this paper could also support remote access to logistics labs and cooperative research scenarios.

As software agents and machine control within one entity are separated and just connected by middleware, it is possible that agents running at the one logistics lab can control a pilot plant at another location via VPN. All real-time communication will still be executed on the machine control layer per fieldbus, but the (non-real-time) functions, such as path planning or order generation/negotiation, can be executed by the remote multi-agent system. This enables a remote testing of the agents' algorithms and of various material flow strategies.

It is also imaginable that pilot plants at different locations "share" their software agents, so that they can be virtually "merged". Therefore, the agents of the foreign plant are used to generate emulator agents to expand the own pilot plant (as described in section 3.2). This would allow the implementation of extensive test scenarios with comparatively little effort. This vision requires a close cooperation between logistics labs, especially regarding the development of compatible software agents and their interfaces to machine control.

References

1. Chisu, R.: Kommunikations- und Steuerungsstrategien für das Internet der Dinge. Dissertation. Lehrstuhl fml, Technische Universität München, Garching (2010)
2. Günthner, W.A., Reinhart, G. (eds.): Abschlussbericht MATVAR – Materialflusssysteme für variable Fertigungssegmente im dynamischen Produktionsumfeld. Herbert Utz Verlag Wissenschaft, München (2000)
3. Günthner, W.A., Wilke, M.: M2 – Wandelbare Logistikstrukturen in Minifabriken. In: Lindemann, U. (ed.) Arbeits- und Ergebnisbericht des Sonderforschungsbereichs 582 der Technischen Universität München, München, pp. 299–333 (2004)
4. Günthner, W.A. (ed.), Heinecker, M., Wilke, M.: Abschlussbericht AiF-FV-Nr. 14021 N: Modulare Materialflusssysteme für wandelbare Fabrikstrukturen. Lehrstuhl fml, Technische Universität München (2006)
5. Günthner, W.A., ten Hompel, M. (eds.): Internet der Dinge in der Intralogistik. Springer, Heidelberg (2010)
6. Kuzmany, F.: Konzeption und Entwicklung von Modulen für das Internet der Dinge. Dissertation. Lehrstuhl fml, Technische Universität München. Garching (2010)
7. Corkill, D.: Blackboard Systems. AI Expert 6, 40–47 (1991)
8. JADE – Java Agent Development Framework, `http://jade.tilab.com`
9. Dijkstra, E.W.: A note on two problems in connexion with graphs. Numer. Math. 1, 269–271 (1959)

Combining Machine-to-Machine Communications with Intelligent Objects in Logistics

Javier Palafox-Albarran[1,2], Alexander Dannies[1], Bala Krishna Sanjeeva[1], Walter Lang[1], and Reiner Jedermann[1]

[1] Institute for Microsensors, -actuators and -systems (IMSAS)
University of Bremen, Otto-Hahn-Allee NW1, Bremen, Germany
{jpalafox,adannies,bsanjeev, wlang,rjedermann}@imsas.uni-bremen.de
[2] International Graduate School for Dynamics in Logistics
University of Bremen, Hochschulring 20, Bremen, Germany
jpalafox@imsas.uni-bremen.de

Abstract. The combination of Machine-to-Machine (M2M) communication and intelligent objects can largely improve the supervision of logistic processes. This requires global mobile communications and short-range wireless sensor nodes. We assembled a demonstrator using off-the-shelf hardware for tests and classroom experiments that shows the feasibility of implementation of the future Internet of Things. It includes the use of embedded devices to perform local intelligent data processing and allows an evaluation of the advantages of ubiquitous M2M communication on a laboratory scale. New software features can be deployed, either to the gateway-device in the form of OSGi-bundles or to the sensor nodes in the form of MIDlet-suites, by using M2M-technology. As an example, we programmed an algorithm for predicting temperature curves in a container using real acquired datasets. A gateway bridges the local and the global network. Sensor messages can be forwarded via email and SMS or be provided by a web server.

Keywords: Internet of Things, Machine-to-Machine Communications, Ubiquitous Computing, Wireless Sensor Networks.

1 Introduction

Machine-to-Machine communication (M2M) is the automatic communication between machines without human interaction allowing direct access to real-time data. One example is a smart meter, which measures the consumption of electricity or water at home or in a factory. It transfers the readings to a remote central system on a daily or hourly basis. The remote system is able to review or act on the collected data. The communication system for transferring data from the remote facility may be, for example, the cellular network and communication satellites. The selection depends on the cost and connection availability. Satellite is the most expensive solution, but is often the only solution in very remote areas or overseas. The solution of cellular

Dieter Uckelmann et al. (Eds.): ImViReLL 2012, CCIS 282, pp. 102–112, 2012.

networks is gaining much attention as it usually fits the best; due to dropping costs of sending data and their wide spread coverage.

Wireless Sensor Networks (WSN), on the other hand, is an emerging technology to monitor ambient conditions. However, they are commonly considered to be stand-alone; the sensor nodes communicate with other sensors and the gateway, but are in principle unable to communicate with the outside world. To change that, the Open Geospatial Consortium [1], is making an effort to establish open geospatial and location standards for exploiting Web-connected sensors and sensor systems.

The combination of M2M and WSN technologies brings the opportunity to profit from their respective advantages. Real world data, collected in the environment, can be integrated into the information world by sending data to the Internet. Furthermore, the system may become ubiquitous, which conceptually means "existing or being everywhere, especially at the same time".

The research is in an early phase and addresses the integration of both technologies into logistics separately. In the BRIDGE project [2] the existing Electronic Product Code (EPC) network [3] was integrated with the Open Geospatial Consortium (OGC) sensor web enablement architecture [1] to show the benefits of the integration of sensor data in logistics. In [4] Kim and Ryu propose an extension of an RFID middleware for active sensor tags. Aberer in [5] and Jajekyu in [6] also combine concepts of object identification, sensor data and the Internet. In a commercial system called Smart Trace [7] M2M is used as an online cold chain monitoring system where temperature changes inside the container are measured by wireless smart tags and uploaded to a server.

While the focus is put on improving the monitoring of the assets through cold chain, intelligent data processing and remote deployment of new software components are not considered in the mentioned approaches. Information about the cargo, provided by intelligent algorithms, will help a human operator or a machine to take efficient logistics decisions. If such algorithms are required, they must be deployed remotely as access to the cargo is often impossible through the transportation.

This paper presents a lab-demonstrator developed by the Institute for Microsensors, -actuators and -systems (IMSAS) as part of the "Intelligent Container Project" by using only open software. Its intended purpose is using it as a tool for educators, students and researchers to evaluate the impact of M2M communication in food transportation supervision.

As the quality of goods mainly depends on the temperature during transportation, this environmental parameter is chosen for demonstration. It shows the best of both technologies. To demonstrate how sensor nodes communicate between each other and process sensor data locally, an algorithm was implemented, which foresees the final temperature at the end of transport in a refrigerated container transporting perishable goods. The algorithm applies the Feedback-Hammerstein training and prediction algorithms [8] to estimate the parameters of a model describing the temperature changes. The sensors send the actual and future temperature via radio to a gateway which generates a notification either via SMS or e-mail to inform about the current and future conditions.

In most wireless monitoring applications the physical access to the sensor is not an option. Nevertheless, it may be necessary to change the software for data acquisition or data processing during the life-time of the sensor. The demonstrator shows how this challenge can be handled: new software components can be uploaded remotely to the gateway as well as to the sensor nodes.

2 State of the Art

It is hard to find a definition that clearly distinguishes between M2M and WSN technologies. Both technologies can utilize sensors to perform remote monitoring and communicate with each other through wireless communication. Knowing the subtle differences will help to understand the implications of their combined use in logistics.

M2M is a technology that allows communications-enabled remote devices to exchange information automatically without human interaction. The basic four stages according to [9] are: data acquisition or collection, transmission of data through a communication network, assessment of data, and response to available information. In general, local devices with missing or only limited intelligence and computing capabilities are used for data acquisition. Their task is to transmit data periodically or to send alarms when a threshold is exceeded or a malfunction is detected. Data is sent to the telemetric system via wired or wireless communication. The assessment of data and the response to it are made through human interaction due to the fact that the local devices are unable to do it autonomously.

WSN, on the other hand, is an emerging technology to monitor ambient conditions. However, they are commonly considered to be stand-alone; the sensor nodes communicate with other sensors and the gateway, but are in principle unable to communicate with the outside world.

WSNs and wireless M2M devices are equipped with sensors, a radio transceiver, an antenna, processing capabilities and an energy source. However, there are differences: wireless M2M covers applications involving longer range and the node will typically be powered from the machine itself; WSNs consist of several sensors interconnected, usually powered by batteries and cover applications involving shorter range.

M2M are deployed when power consumption is not critical, the size/weight of the devices is not an important factor and a range of kilometres is required. Additional features may include, for example, bidirectional communication.

WSNs are to be deployed in short/medium range areas where human intervention is not possible – either because it is too dangerous, like in a battlefield or a forest fire, or too remote to send people, as is the case when monitoring glaciers and mountains. The system must replace human intervention with spatially distributed sensor nodes. The desired characteristics of an autonomous WSN deployment include high lifetime and robustness as well as fault tolerance and self-configuration [10]. Furthermore, it cooperatively monitors physical or environmental conditions, such as temperature, humidity and pressure, at different locations.

2.1 The SunSPOT as Example Sensor Node

There are several sensor node platforms on the market today available for researchers and students. The most frequently used are TelosB, programmed using a C-like language and TinyOS, and IMote2, which is highly configurable and can run the operating system Linux, with a Java Virtual Machine (JVM) and an OSGi-framework on top of it. However, learning the configuration and getting used to both of them is too time consuming to be used in a demonstrator.

The chosen platform to be used in the demonstrator is the Oracle SunSPOT [11], because both the hard- and software are open source. It is possible to write applications in Java (Java ME), which are platform-independent. Using an Integrated Development Environment (IDE) like NetBeans makes the deployment (with the internal use of ANT-scripts) of new applications very easy.

The core of the SunSPOT is an Atmel AT91SAM9G20 processor running at 400 MHz, integrated with a multichip package consisting of 8MB Flash Memory and a 1MB SRAM memory, power management IC to go to deep-sleep mode, an IEEE 802.15.4 CC2420 chip and an inverted-F antenna printed on the circuit board. The interconnected sensors are: a temperature sensor integrated in the ADC LTC2487, a three axis accelerometer MMA7455L, and an ADJD-S311-CR999 RGB light sensor [8].

The Squawk Java VM is pre-installed on the SunSpot. Squawk is targeted for small resource constrained devices. It utilizes the concept of isolates, where several applications running on the VM can share common suites, leading to a significantly reduced memory footprint and multi-threading. The used Connected Limited Device Configuration (CLDC) only contains the minimum amount of classes necessary to enable the operation of a JVM. The Mobile Information Device Profile (MIDP) allows applications written, the so-called MIDlets, to be used, installed and removed dynamically.

3 The Impact of Combining M2M with WSN in Logistic Practice

In conventional transport logistics it is required to have the right product, in the right quantity, at the right place at the right time. In this context, an instant identification of the assets in the supply chain is helpful, which is possible by reading RFID tags.

It is important to point out that when considering perishable goods, the data management becomes dynamic and the dependence on the previously mentioned requirements becomes stronger. The increase of such dynamic behaviour is conditioned by the quality data, which determines the dynamic of the other aspects. For instance, the right time, quantity and place may vary during the transportation. As the quality of the goods decreases, new routes and suppliers/buyers have to be found, according to the actual price of the cargo, with the aim of increasing the profitability.

At the same time, this quality data is susceptible to fluctuations in environmental parameters, such as relative humidity or temperature. The deteriorations can lead to a decrease in the aesthetic appeal, as well as a reduction in nutritional value. The information about the quality/temperature of the product must be available at any time and everywhere, in order to have valuable information that allows taking proper decisions. As a result, the amount of goods that arrives at the costumer in non-acceptable conditions decreases. This leads to further advantages, such as reduction of transport volume and greenhouse gas emissions. Actions against faulty cooling conditions can be taken as soon as a problem arises. Goods can be sorted in the warehouse by their actual quality condition. It requires the deployment of more sophisticated technologies to include the mentioned dynamic fluctuations in the supply chain.

Combining M2M and WSN in logistic practice has the potential benefit of waste reduction, which is achieved by a greater amount of and more accurate just-in-time information. Well accepted global technical standards, such as EPCglobal [3], that enable instant identification of items in the supply chain may be extended to include management of sensor data. In EPCglobal real objects are linked to the Internet through the use of the unique electronic product code (EPC), such representation of objects in the network is attributable to the Auto-ID centre [12, 13] and is called "The Internet of Things" (IoT).

The EPC standard is based on static data and was conceived as an efficient tool for the traceability of the assets. The cold chain, which is a special case of supply chain, requires, in addition to traceability, environmental data monitoring to provide complete information about quality of the goods. The temperature of frozen and chilled goods must be monitored continuously throughout the entire cold chain.

Uckelmann, Harrison and Michahelles [14] defined the Internet of Things from a logistic point of view. As they mention, information about the right condition and at the right price have to be included in addition to traceability. They describe the requirements for the future Internet of Things. In a broader vision, the possibility of including embedded devices such as WSN is mentioned.

M2M and WSN will fill the gap between autonomous logistics and the Internet of Things. M2M will allow us to access information in a ubiquitous way at a reasonable price by sending the right e-mail or SMS with the condition information to be readable for human interaction. WSN may use sensor information to detect signs of degradation of condition at an early stage with the help of intelligent data processing.

4 Concept of the Demonstrator

Fig. 1 displays the concept of the demonstrator. It consists of three categories: WSN, gateway and end-user; in each one of them only open software is used. Conventional remote monitoring is possible by using M2M communication (which is represented by dashed lines in the figure), if the sensor nodes are programmed to transfer the readings periodically to be visualized by the end-user and/or to send and alarm when a temperature threshold is exceeded. However, such a solution may lead to unnecessary data transmission. Our implementation differs to that in running algorithms on the sensor boards to predict the temperature change and shelf life. The task of the WSN is the gathering of environmental data – here temperature values – and local data-processing. The processed data is transferred wirelessly and can be received by the base station – the juncture between the WSN and the gateway.

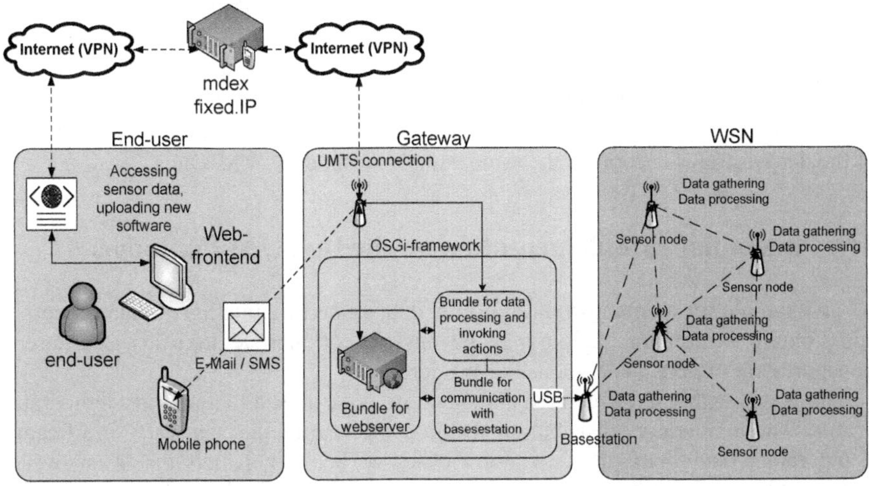

Fig. 1. Concept of the demonstrator (M2M communications represented by dashed lines)

Through the gateway it becomes possible to connect the sensor world with the information world. For this purpose, the Equinox OSGi-framework is installed on top of the Linux OS, which enables a high degree of dynamics by allowing the installing or updating of software modules remotely during runtime. An OSGi-bundle contains a web server and servlets for generating dynamic web-pages – this is for displaying data and altering of software on the sensors. Furthermore, another bundle encloses an implementation for sending notifications via SMS or e-mail. The end-user can decide, if he is notified of selected events, for instance exceedance of a specified temperature threshold; can take a look at gathered data of the WSN; or even has the ability to deploy new software to a chosen sensor node.

From the different types of wireless sensor modules, we selected the SunSPOT because it provides the option to update or deploy new Java applications without physical access to the sensor. One or more applications, so called MIDlets, can be combined to a MIDlet suite and transferred over a radio link to the sensor node.

The currently available example applications for managing a network of SunSPOTs are based on a PC providing graphical interfaces and a human operator. The idea of M2M Deployment goes beyond this concept. Firstly, it should be possible to deploy MIDlets in remote sensor networks without a fixed connection to the internet. In our solution the sensor network is linked by a gateway to a GPRS or UMTS cellular network for global communication. The gateway has only limited recourses compared to a PC and should operate without any local user interface. Secondly, we pursue the vision that machines deploy updated software on other machines. Therefore, the deployment process must be fully automated and executable over multi-modal networks. A backend machine, for example a central server for environmental supervision, decides when it is necessary to update the software of some sensors in the field to perform a new measurement or data evaluation task.

Instead of using a telematics unit with a base station, a GPS/GPRS/UMTS module can directly be connected to the SunSPOT. If the module has very low power consumption, the powering eventually becomes possible via the SunSPOT itself. Otherwise, an external power source for supplying the SunSPOT and this communication module would be necessary. This solution allows all basic functions of the demonstrator except for the remote deployment of new MIDlets.

5 Exemplary Implementation on the Involved Platforms

As an example for a demonstration of how data gathering, intelligent data processing and ubiquitous communication may be applied to improve logistics, the case of transportation of perishable goods is considered.

The goods within a refrigerated container are not exposed to a uniform temperature profile. The temperature distribution within a single container can vary significantly. Temperature deviations of ±2 °C were observed between pallets inside an oversea container loaded with bananas [15]. The temperature profile depends on the ambient temperature, the total air circulation rate and distribution, external environmental conditions, and the respiration heat of the goods. Packing mistakes can either block or short-circuit the circulation of cooling air. In this case, temperature deviations can rise to up to 10 °C.

Sensor nodes are to be co-located within the goods as depicted in Fig. 2. The data of the air supply and on the boxes containing the goods is gathered and processed by the sensor nodes. The processing in this example is made by the node in the box by calculating the parameters of an equivalent model of the refrigerated container. Those parameters are transferred between the sensor node and the gateway. The gateway may send it to a remote server, which can use them to predict future temperatures.

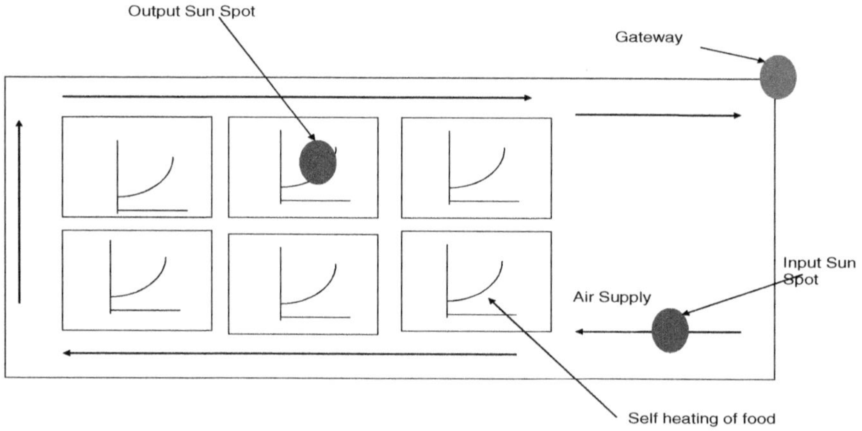

Fig. 2. Implementation of the platforms in a refrigerated container

5.1 Data Gathering

The SunSPOT located near the cold air supply samples the local temperature periodically and broadcasts it to the rest of the sensors located inside the boxes. These sample the local temperature in the boxes every time a measurement from the air supply sensor arrives. For demonstration purposes, the reading from the sensors is replaced by reading an array of floating values containing the resulting datasets from an experiment during a shipment of bananas from Costa Rica to Antwerp in May 2008 [15].

5.2 Data Processing

The future temperature values inside the container can be calculated by using system identification techniques, which estimate the missing parameters for a given model structure. A Single-Input Single-Output (SISO) grey-box model is used to predict the temperature inside the container under the presence of perishable goods with the aim of reducing the complexity and preserving the accuracy. The proposed, so called Feedback-Hammerstein with white noise model, is shown in Fig. 3. It provides a meaningful description of the factors involved in the physical system including the effect of transporting living goods, such as fruits and vegetables.

In order to perform an efficient data processing on the SunSPOT, an online recursive method was chosen, as it requires much lower resources in terms of memory and CPU power than offline counterparts. The difference equation of the model presented in Fig. 3 is described by equation 1.

$$a_1^*(q^{-1})y(t) = b_1 \alpha u(t) + b_1 \beta (e^{\gamma y(t)} - \gamma y(t)) \tag{1}$$

In the model, γ is a key parameter that characterizes the heat production in Watts per kg of fruit as a function of temperature. The value of γ depends on the type of fruit, but is constant for all products of the same type. The following parameters have to be estimated: β, which is a scaling factor depending on the amount of food and is given in kilograms, b_1, which is the zero of the first-order linear system and a_1^*, which is the pole of an equivalent pseudo-linear system. The algorithms to calculate the required parameters of the resulting system are presented in [8]; they have lower order matrix dimensions and do not need any matrix inversion. In total, three parameters for the equivalent system are estimated ($a_1^*, b_1\alpha$ *and* $b_1\beta$) and are updated after each measurement. The Java-math-library is still not available on JavaME. Required mathematical tools, such as matrix inversion and exponential function, were implemented and tested as additional features and for further expansions of the algorithm . In order to give an accurate prediction, the model parameters have to be iterated over three days at a measurement interval of one hour, equivalent to 72 cycles.

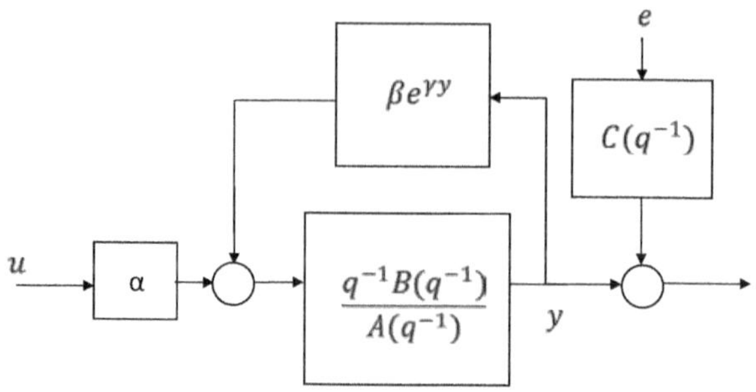

Fig. 3. Feedback-Hammerstein Model

5.3 Data Forwarding

After the data processing in each SunSPOT is performed, the three resulting model parameters and the last supply and output parameters are forwarded to the base station. The base station is connected to the gateway device which provides an OSGi-framework. Different software modules are installed in this environment. One bundle receives data from the base station and also runs a prediction algorithm, which is able to calculate each point of the output temperature profile for the remaining transport duration – typically two weeks. Based on the calculations in correspondence with a defined threshold value, an event can be triggered in the OSGi-context. To be able to react on these events, additional bundles can be installed. One application, which can be connected to the environment, is a web interface for displaying data in form of a table or graph. It can also be used for remote configuration, for instance to change the threshold value for notifications. Other pieces of software can be added to the environment, which send notifications via SMS or e-mail when receiving an event. For sending these notifications, an uplink to a mobile service provider is necessary. For being able to access the provided web interface from around the world, a two-way communication needs to be established. Due to the fact that mobile service providers only assign IPs in the private network address range like 10.x.x.x, a connection to the gateway device can't be established. An extra service is needed to allow that. Here, we choose the gateway provider mdex, which represents the juncture between the two ends: WSN-gateway and end-user. Both ends join a virtual private network at the service and so are able to communicate with each other.

5.4 Deployment of MIDlets over M2M Communication

For enabling an upgrade of the intelligence of the sensors in the WSN, a mechanism for the deployment of new software is necessary. The solution in this context consists of three steps. Firstly, the new Java code has to be compiled and linked, which results in a jar-archive, which is then used to create a MIDlet suite. In the second step, this MIDlet suite has to be transferred to the gateway device, which is connected to the base station. This can be done via an upload dialogue provided by the web interface.

The final step is the execution of a deployment script, which is also accessible via the web-frontend. This script installs the suite on the SunSPOT sensor, which completes the upgrade process.

6 Evaluation and Discussion

Traditionally, logistics is based on traceability for static data. An ID tag is assigned to the goods to know if they are going to arrive to the designated place in the right quantity and time. EPCglobal standards manage such information globally, basically based on RFID tags.

Recently, the necessity of much more interactive and intelligent features has been recognized. They should allow an efficient management of dynamic data as required, for example, during the transport of perishable goods. Sensing, interaction with the environment, mobility and local intelligence are trends for the future Internet of Things, which will optimize the information flow.

The combination of the respective advantages of the technologies RFID, WSN and M2M communication in transport logistics explores a new technical horizon. This leads to a breakthrough in the future Internet of Things, which will allow a global and pervasive traceability of the location and quality state of the assets.

This paper is presented with the aid of a demonstrator showing how the pervasiveness of M2M combined with the embedded intelligence of wireless sensor technologies will impact the management of the dynamic behaviour on food transportation.

Feedback-Hammerstein training and prediction algorithms were used as an example of the intelligent features that can be deployed in wireless sensor nodes. M2M was demonstrated by sending the prediction results from the sensor node to a telemetric system that sends an SMS or an e-mail to be read by a human operator.

The demonstrator was implemented completely using free software. Linux and the OSGi-framework Equinox, with the additional bundles needed in this context, were installed on the Gateway, and Java ME was used for the SunSPOTs. It also demonstrates additional advantages of dynamic updates supported by Java and OSGi by allowing MIDlets that may contain intelligent algorithms required to deal with the actual cargo condition to be deployed remotely onto the sensors in real time.

Acknowledgements. The research project "The Intelligent Container" is supported by the Federal Ministry of Education and Research, Germany, under reference number 01IA10001, and by the German Research Foundation (DFG) as part of the Collaborative Research Centre 637 "Autonomous Cooperating Logistic Processes". The current study is also supported by International Graduate School in Dynamics in logistics at Bremen University. Further information about the project can be found at http://www.intelligentcontainer.com.

References

1. Open Geospatial Consortium, http://www.opengeospatial.org/
2. Bowman, P., Ng, J., Harrison, M., López, T.S., Illic, A.: Sensor based condition monitoring. BRIDGE (2009), http://www.bridge-project.eu/data/File/BRIDGE_WP03_sensor_based_condition_monitoring.pdf
3. EPCglobal, http://www.gs1.org/epcglobal
4. Kim, H., Ryu, W., Hong, B.: Extension of RFID Middleware Platform for Handling Active Sensor Tags. In: Proceedings of the 2010 Fourth International Conference on Sensor Technologies and Applications, pp. 163–168. IEEE Computer Society (2010)
5. Aberer, K., Hauswirth, M., Salehi, A.: Global Sensor Networks, Technical report LSIRREPORT- 2006-001 (2006), http://lsirpeople.epfl.ch/salehi/papers/LSIR-REPORT-2006-001.pdf
6. Jaekyu, C., Yoonbo, S., Taekyoung, K., Yanghee, C.: SARIF: A novel framework for integrating wireless sensor and RFID networks. IEEE Wirel. Commun. 14(6), 50–56 (2007)
7. Smart Trace, http://www.smart-trace.com/index.php
8. Palafox-Albarrán, J., Jedermann, R., Lang, W.: Energy-Efficient Parameter Adaptation and Prediction Algorithms for the Estimation of Temperature Development Inside a Food Container. In: Cetto, J.A., Ferrier, J.-L., Filipe, J. (eds.) Informatics in Control, Automation and Robotics, pp. 77–90. Springer, Berlin (2011)
9. M2M Communications, http://www.m2mcomm.com/about/what-is-m2m/index.html
10. Verdone, R.: Wireless sensor and actuator networks: technologies, analysis and design. Academic Press (2008)
11. Sun Labs.: Sun™ Small Programmable Object Technology (Sun SPOT) Theory of Operation (2007), http://sunspotworld.com/docs/Purple/SunSPOT-TheoryOfOperation.pdf
12. Haller, S., Hodges, S.: White Paper: The Need for a Universal SmartSensor Network. Auto-ID Centre Institute for Manufacturing, University of Cambridge (2002)
13. Mitsugi, J., Inaba, T., Pátkai, B., Theodorou, L., Sung, J., Sánchez López, T., Kim, D., McFarlane, D., Hada, H., Kawakita, Y., Osaka, K., Nakamura, O.: Architecture Development for Sensor Integration in the EPCglobal Network. Auto-ID Centre Institute for Manufacturing, University of Cambridge (2007)
14. Uckelmann, D., Harrison, M., Michahelles, F.: An Architectural Approach Towards the Future Internet of Things. In: Uckelmann, D., Harrison, M., Michahelles, F. (eds.) Architecting the Internet of Things, pp. 1–24. Springer, Berlin (2011)
15. Jedermann, R., Moehrke, A., Lang, W.: Supervision of banana transport by the intelligent container. In: Kreyenschmidt, J. (ed.) 4th International Workshop, Coolchain-Management, pp. 75–84. University Bonn, Bonn (2010)

Augmented Traceability with Internet of Things

A Demonstrator for Seafood Supply Chains

Nicolas Pauvre[1], Jacques Madelaine[2], Jérôme Le Moulec[2], and Adrien Laurence[2]

[1] GS1 France, 2 rue Maurice Hartmann, 92137 Issy-les-Moulineaux, France
Nicolas.Pauvre@gs1fr.org
[2] GREYC - CNRS UMR 6072, Bd Maréchal Juin, F-14000 Caen, France
{Jacques.Madelaine,Jerome.LeMoulec,
Adrien.Laurence}@info.unicaen.fr

Abstract. Imagine seafood supply chains where products tell you when and where they have been caught. Through graphical interfaces and applications running in live mode, the demonstrator presents a standard-based traceability system for seafood supply chains that increases business benefits and enhances consumer experience. The demonstrator stems from the research project "WINGS"[1] that aims at designing architectures embedding Discovery Services to retrieve decentralized data warehouses of manufacturing goods. The French Research Agency (ANR) [2] supports WINGS in the framework of the program "Future Networks and Services" (VERSO).

Keywords: Demonstrator, EPCIS, EPCglobal, GS1, ONS, DNS, Discovery Services, Standards, Supply Chains.

1 Introduction

Global supply chains operating in a networked business-to-business world present tremendous challenges and opportunities for the so-called "Internet of Things". In prospect, huge amount of different types of objects connected to seamless and ubiquitous networks will bring end-to-end visibility to the supply chain. To do so, interoperability between enterprise information systems is an essential condition, as outlined in the Global Commerce Initiative's Report – 2016: The Future Value Chain [3].

Since the launch of the EPCglobal standards, more and more companies have started to explore the possibilities of using the related technologies, services and interfaces, such as the EPCIS [4], which represents the first step towards the usage of the EPCglobal network. The foundation for the kind of connectivity will increase visibility throughout global supply chains with a various class of new applications such as helping trace shipments, combat the introduction of counterfeited products and prevent retailer's out-of-stocks.

To illustrate this next phase of supply chain development, GS1 and GREYC have designed and implemented a standard-based platform that simulates seafood supply chains as described in Fig. 1.

Dieter Uckelmann et al. (Eds.): ImViReLL 2012, CCIS 282, pp. 113–121, 2012.

One of its key standard-based components is an object directory service, called ONS (Object Naming Service) [4], which provides a lookup mechanism that enables anyone to find the address of services, such as traceability repositories (EPCIS). The ONS implemented in the platform is based on the proposition of the WINGS project in order to benefit of a multi-rooted naming service. With the same goals, the platform functions in interaction with Discovery Services (DS) to allow access to all the data warehouses along the supply chain. The DS are mandatory when the supply chains are not simple linear chains as in our use case. Finally, a component at the application level is used to display the supply chain using an interactive map.

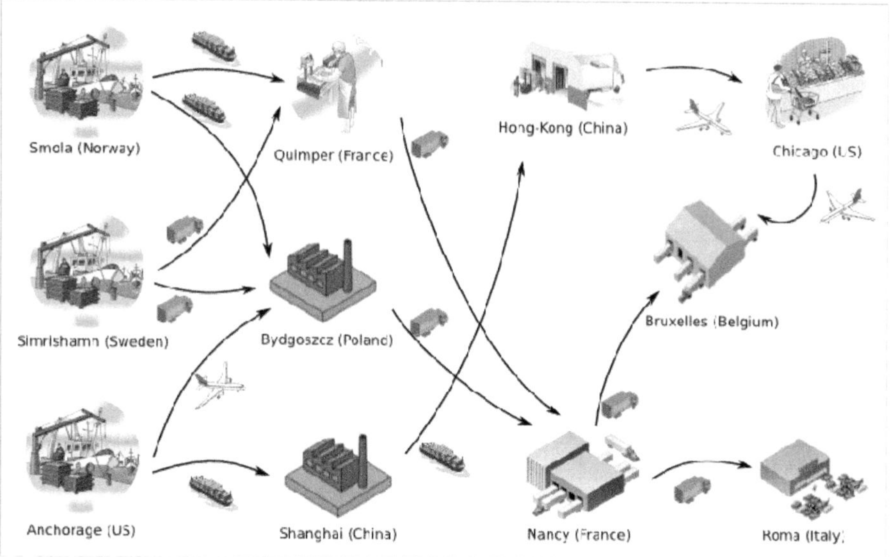

Fig. 1. Seafood supply chains

2 Architecture and Components of the Demonstrator

The demonstrator is installed on a distributed platform, the so-called "IoTa platform"[1]. IoTa gathers nine EPCIS, three DS and the ONS system. These services are distributed over three locations (the GREYC laboratory, the CERTIC from the university of Caen Basse-Normandie and Orange Labs Caen). Fig. 2 shows the deployment of these components. This platform puts the EPCglobal components in a realistic situation according to the network heterogeneity and its inherent problems: firewalls, links speed, etc. Only the EPCglobal upper layers are deployed in IoTa. The components located in the capture layer (readers and ALE) are simulated using a Petri net simulator.

---

[1] There is no direct connection with the European project so called IoT-A; `http://www.iot-a.eu/public`

The ONS system is used to locate the information system (the so called EPCIS) that stores the different events reflecting the objects along the supply chain. But there are several EPCIS used to store events about the same object (same identifier or EPC) as the supply chain involving many partners. In order to build a distributed architecture, ONS is used to locate the first EPCIS, while DS allow giving access to all the EPCIS storing events about that object. The platform allows DS interconnection as described in [4], to benefit from a true distributed architecture.

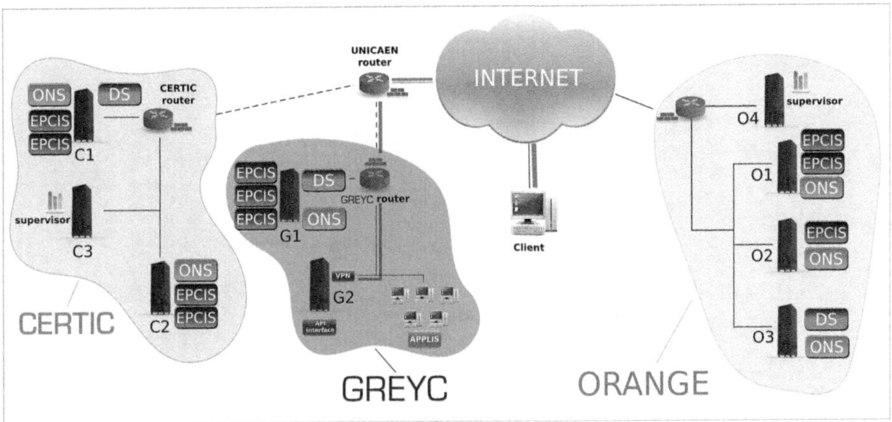

Fig. 2. The IoTa Architecture

The platform hosts two supervisors for servers and network monitoring. Thus, it can be used to experiment various solutions and measure their costs in terms of bandwidth and response time.

This section describes the different components of IoTa, starting from the lower layers components.

2.1 The Capture Layer Simulator

The demonstrator simulates the readers and all the components of the capture and filtering layer using a Petri net. The simulator is based on the kernel called PNK-2.0 developed at the University of Trier [6].

In the model, transitions are used to represent the readers and the components of the capture and filtering layer. Places represent warehouses and transports. Thus, the supply chain depicted in Fig. 1 is simulated with a Petri net presented in Fig. 3.

With the Petri net kernel, call back functions triggered when a transition fires, can easily be defined. In our case, an EPC event is synthesized and sent to a corresponding EPCIS. When a transition has several input places, the transition fire produces an aggregation event. In that case, there is always a printer among the input places that makes the new tag.

Furthermore, the kernel has been enhanced in a few ways. The graphical user interface is more friendly. It is now possible to figure the places and the transition using any image (e.g., warehouse, factory, boats...). Popup menus may be used to parametrize places and transitions. In order to make realistic simulations, transitions fire time intervals can be defined.

The Petri net model has one restriction: at most, one transition may fire at a given time. This makes simultaneous event generation impossible. To overcome this drawback, it is possible to connect several simulators using *pipes* in order to simulate more exactly real environments.

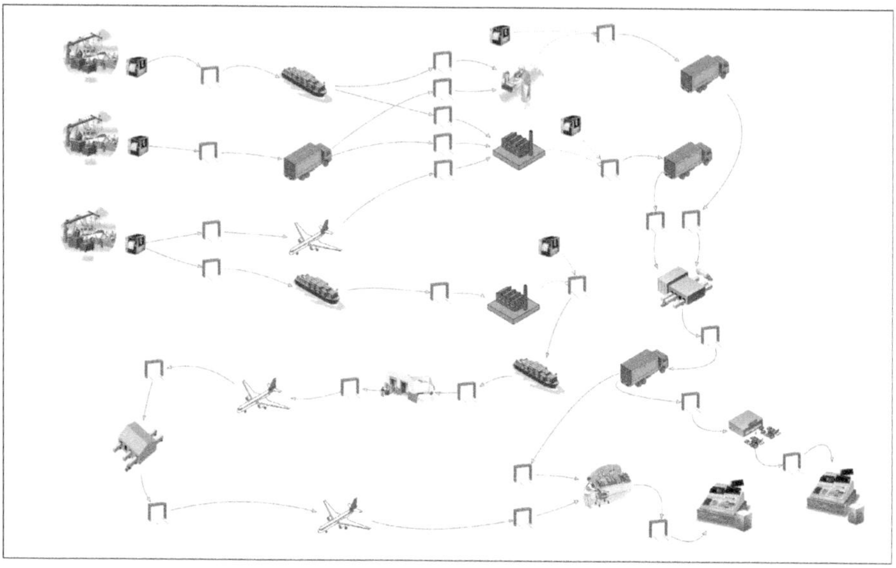

Fig. 3. Petri net model for the seafood supply chains

2.2 EPCIS

IoTa uses the Fosstrak [7] implementation of the EPCIS standard. It offers two interfaces. The capture interface over HTTP is rather crude in the sense that a new event declaration is made with a simple HTTP/GET request. This event is stored in the local EPCIS repository. As it offers more complex requests and replies, the query interface uses the SAOP/XML protocol over HTTP.

The EPCIS also offers a publish/subscribe mechanism. As the platform IoTa has DS servers, events captured by the EPCIS must trigger publications to the DS in order to be indexed. In IoTa, we chose to realize the publication to the DS using this publish/subscribe mechanism. It has the main advantage to need no modification to the EPCIS Fosstrak implementation. A new separate component, called IS2DS, has been designed. At the application deployment, it subscribes to the EPCIS for every new capture. Every minute, it receives a bunch of new inserted events. They may then be forwarded to the DS using the ESDS-1.0 protocol.

2.3 Discovery Services (DS)

The DS standards have been implemented by the GREYC laboratory. The DS servers receive information from the EPCIS, so that they hold information for a given EPC code as soon as a new event is inserted. The DS server can then be deployed by users to find which EPCIS to query to find information about events concerning that given code.

In order to allow several DS servers, we must have a mechanism that avoids querying every DS to discover every EPCIS that keeps track of a given EPC code. The DS servers in IoTa implement the solution called "DSs" as explained in [5]. With that solution, a dedicated DS is chosen to be the so-called *referent* DS. This DS index all the DS as well as the multiple EPCIS associated with the EPC code. Whenever a user's level application wants to trace an object knowing its EPC code, it has to query the referent DS for the DS. These DSs will be in turn query for the EPCIS that will give all the relevant events. The ONS has been enhanced in order to keep track for the referent DS.

2.4 Object Naming Service (ONS)

The ONS standard is inspired from the Internet Domain Name System (DNS). Like the DNS, the ONS servers stand in a hierarchy of servers, but instead of associate an IP address to an Internet name, they associate an EPCIS IP address to an EPC code.

For IoTa, as explained in the previous section, ONS stores also the referent DS address for any EPC code. The DS address is stored in a NAPTR record under the qualification EPC+DS instead of EPC+EPCIS for an EPCIS address.

Furthermore, the ONS used is multi-rooted. Two versions are available: FONS [8] and PRONS. With FONS, root servers are dispatched using the country part of the EPC code. All the root servers share an associative array mapping country code to root server addresses. When a local ONS must answer to a query for a code it does not know, it forwards that query to its upper level server (just as for the DNS). If the root server still can't answer, it uses the shared array to query the right root server that will be able to redirect the query to the correct local ONS. With the PRONS solution, the root servers share the information about the localization of a code implementing a dynamic distributed table.

2.5 Application Layer

The application layer gathers all the height level user's applications. In order to simplify the development of such applications, the IoTa platform offers a dedicated interface simply named the "application interface".

The "application interface" knows about all the EPCglobal components client interfaces. It is able to query the different components of the IoTa platform: EPCIS, DS and ONS as shown in Fig. 4. Therefore, any client application that needs to access a resource stored in IoTa may use this interface. The "application interface" offers two different protocols for the query interfaces: RMI and SAOP/XML. It can be considered as the IoTa API. Several high level methods are available such as queryEPCIS, queryDS or traceEPC.

Let's illustrate its use with an example of a client application that needs to track a given object (given its EPC). The client application must simply call the traceEPC method of the application interface. It is the application interface that will take charge of querying the different components of the platform. It will first use its ONS client module to retrieve the address of the referent DS for the given EPC. It then queries this DS instance using its DS client module to gather the addresses of every DS and EPCIS application that contains information for that given EPC. The application queries, in turn, each DS to retrieve the address of all the EPCIS that holds information on the given code. Finally, as the application interface knows the addresses of every EPCIS storing events about the given EPC, it can retrieve all these events and send them back to the client application as a list of EPC events. This guarantees that the client retrieve a complete list of events to track the object.

The GREYC laboratory has developed several client applications. The first one, the so-called *IoTa validator*, can validate the functionalities of every component. It is used in conjunction with the event simulator. As it knows about the events generated, it can then verify that these events are stored in the correct EPCIS and indexed in a correct DS that may be retrieved thanks to the ONS. Another application is offered to track things.

The next section describes a more visual application that visualizes the route of goods using a cartographic tool.

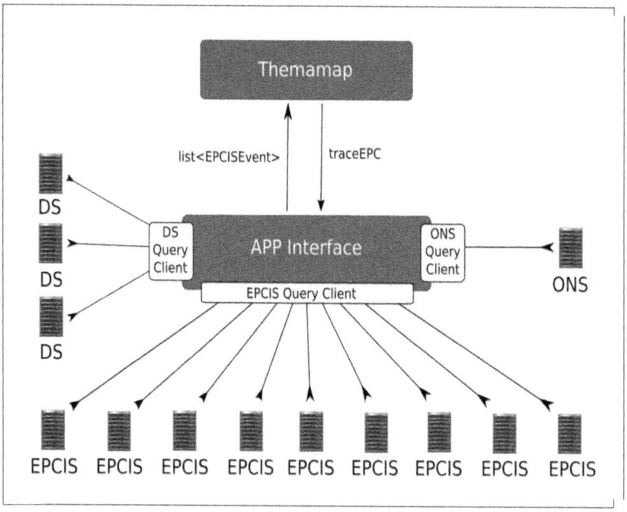

Fig. 4. IoTa application interface and ThemaMap client application

3 Applications of the Demonstrator

Supply chains events, from the fish boat to the retailer, are dynamically stored into the EPCIS servers. The capture of data is generated through the simulator application. Based on the events produced all along the supply chains, a query from an identifier (EPC) leads to the retrieval of data associated with the logistic activities of the seafood product. The traceability data are then displayed into a geographic map, using the ThemaMap application. In that way, the full history of the seafood product life cycle is displayed as shown in Fig. 5.

The product is handled by multiple stakeholders (fishery, processor, distributor, wholesaler, and retailer) in various countries, depending on the business relationship among the stakeholders. Several scenarios of product life cycle are therefore possible.

As an example, the first step is at the boat level, where the raw fish are caught by a fishery from Sweden, Norway or Alaska, and identified with an EPC (commissioning business process).

Raw fish are then packed into logistic units (pallet) for being transported to a fish processor factory located either in France, Poland or in China. As the product is transformed to a new item (fish fillet), a new EPC is commissioned to identify that trade item.

Finally, the fish fillets are bought by a wholesaler or distributor for being sold to consumers in a retail shop in Brussels or in Roma. At point of sale, the product life cycle information are displayed in order to let the consumer know about where the fish is coming from, when it was captured, by who and how, etc. The consumer can also get the traceability information directly on his mobile phone, by scanning the product code.

4 Benefits and Lessons for Supply Chains Management

The demonstrator is used to illustrate a next generation of solutions and services for business intelligence in the supply chain with broadly based and open business enterprise networks. It represents the important move onwards from RFID supply chain applications. The demonstrator figures out what this implies.

Standards implemented enhance data management and quality that is shared all along the supply chains. EPCIS capture and query standard interfaces allow supply chains stakeholders to integrate and aggregate data from different sources in an automatic way. That decreases manual data entry and results therefore in more reliable data to feed the business applications. So the demonstrator proves that a standard-based approach reduces costs by facilitating the delivery of electronic messages to the supply chain partners.

In addition, the flow of data, fully synchronized with the flow of goods in the supply chain, advance the development of innovative logistics services. The ability of ONS to connect users to authoritative sources of information about the seafood product in a standard way, gives rise to a new class of applications offering trusted and value-added information.

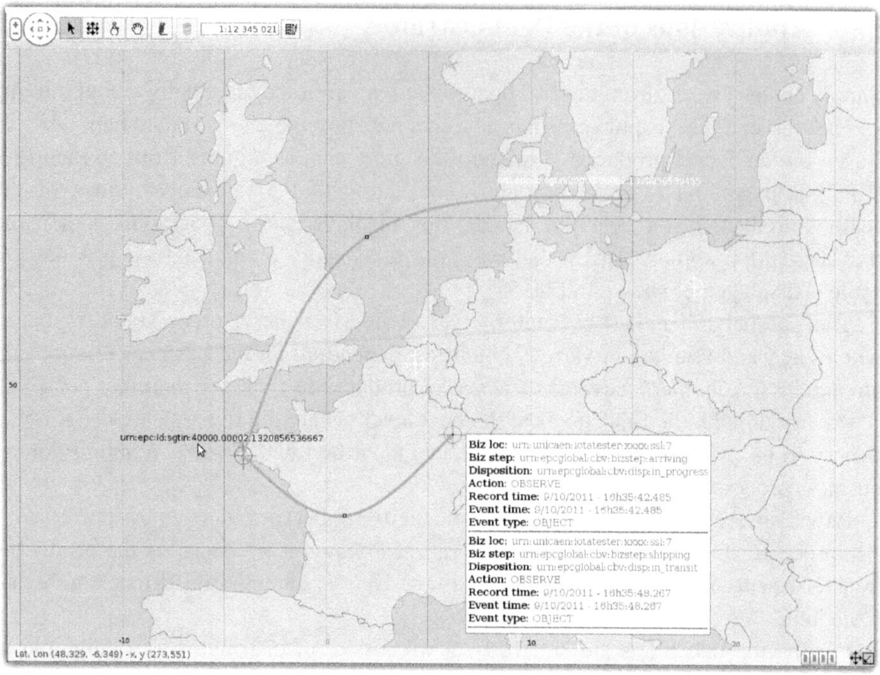

Fig. 5. Traceability of the seafood product presented with the ThemaMap application

The platform enables Internet-based applications to be dynamically developed including – but not restricted to – business-to-business scenarios. As an example, data alignment provides key insights into supply chain activities to serve all corporate functions (marketing, manufacturing, quality, etc.). It could also serve consumers with mobile applications. Through the use of standards from end-to-end supply chains as well as Internet, traceability data on products can be pushed to the consumers. That will increase consumer experience, meet product safety and boost consumer confidence, leading to an increase of sales.

References

1. WINGS Project website, `http://www.wings-project.fr`
2. Global Commerce Initiative's Report: 2016 – The Future Value Chain (2008), `http://www.gcinet.org/gci/content/e29/e1525/e1610/item_d1910/2016_Future_Value_Chain_GCI_Report-06-11-01-ohne.pdf`
3. EPC Information Services (EPCIS) Standard v1.0.1, `http://www.gs1.org/gsmp/kc/epcglobal/epcis/epcis_1_0_1-standard-20070921.pdf`
4. Object Naming Service (ONS) Standard v1.0.1, `http://www.epcglobalinc.org/standards/ons/ons_1_0_1-standard-20080529.pdf`

5. Le Moulec, J., Madelaine, J., Bedini, I.: Discovery services interconnection. In: Proceedings of the 3rd International Workshop on RFID Technology – Concepts, Applications, Challenges (IWRT 2009), pp. 59–68. INSTICC Press, Portugal (2009)
6. Kindler, E., Weber, M.: The Petri Net Kernel: An Infrastructure for Building Petri Net Tools. Softw. Tool. Technol. Transf. 3, 486–497 (1999)
7. Fosstrak: Free and Open Source Software for Track and Trace, `http://www.fosstrak.org`
8. Balakrichenan, S., Kin-Foo, A., Souissi, M.: Federated ONS Architecture for the Internet of Things – A Functional Evaluation. In: Internet of Things 2010 Conference, Tokyo (2010)

From TagID to eVIN – Structured Data Exchange Based on RFID

Bernd Scholz-Reiter and Jeanette Mansfeld

BIBA, Bremer Institut für Produktion und Logistik GmbH at the University of Bremen, Hochschulring 20, 28359 Bremen, Germany
{bsr,man}@biba.uni-bremen.de

Abstract. This paper gives a review about three different lab-based projects. All of them are evaluating the usage of structured data storage based on RFID to optimize data exchange throughout the supply chain. The first project uses ISO standards for data-on-tag storage, while the second one proves the concept of the EPCglobal architecture, which holds data in an EPCIS repository. Both have been implemented in pilot implementations. The third project follows the EPCglobal standards as well. In this case, the lab serves as a place to demonstrate the research concept through a mobile learning game. After a description of each project, the paper offers a comparison of the project objectives, their methods and results as well as the impact of the lab environment.

Keywords: RFID, EPCglobal, ISO, Mobile Game, Distributed Data Exchange.

1 Introduction

Today's logistic systems may be complex and dynamic due to globalization and increasing customer orientation. Therefore, it is necessary to identify and locate objects and enable traceability within the whole logistics system starting from the production processes throughout the supply chain movements. Radio Frequency Identification (RFID) is a capable technique to support the visualization and optimization of logistic processes using automatic identification of objects [1]. It can be used with different goals of process optimization, for instance autonomous control of logistic processes [2] or supporting traceability in reverse logistic supply chains [3]. On the one hand, there is a great variety of the usage of RFID in logistic systems, while, on the other hand, the adaption of RFID into existing processes may cause high costs. Therefore, it seems to be useful to research the range of applications in a lab environment. A lab is an environment that is close to reality and works under predefined conditions. Consequently, there are a lot of possibilities to build realistic scenarios using new or combinations of established technologies.

This paper gives a review of three projects in two lab environments: the LogDynamics Lab at the University of Bremen and the demonstration center of the RFID-based Automotive Network (RAN) project at the Bremen Institute of Production and Logistics (BIBA). All of them deal with the usage of the data stored on a RFID

Dieter Uckelmann et al. (Eds.): ImViReLL 2012, CCIS 282, pp. 122–133, 2012.

transponder using multiple logistics and information technology standards. The following chapters offer a short description of each project. It provides a conclusion evaluating and comparing the impact of the performance of those projects in a lab environment.

2 Data@RFID – ISO Compliant Storage of Data-on-Tag

The project was carried out in the LogDynamics Lab with the goal to utilize the free user memory of a RFID transponder for structured communication. The usage of common Auto-ID techniques like barcodes offers the possibility to store immutable product data. Contrary to that, it is possible to store, change and extend data on RFID transponders. In this way it would be possible to store the information about a product life cycle with the object itself. Data@RFID is a pilot application on a mobile RFID terminal using different standards to store product data in the optional user memory. The application is installed on a NordicID PL3000 device running Windows CE 4.2. The programming language is C# and all logical data about the used standards is mapped to XML files.

The first version of Data@RFID supported Application Identifier, Data Identifier and Text Element Identifier (see Table 1). While the content of the identifier is described in ISO 15418 and ATA Spec 2000, the definition of the data structure is defined in ISO 15434. This standard describes the message header and trailer, format indicators and data separators. The identifiers are stored in the user memory of a RFID transponder using this standard to offer them to all other applications using the ISO 15434. All information about the standards is stored in XML files with the goal to make them changeable and extensible with other standards. Here is an example for the definition of the Application Identifier (referring to the screenshots in Fig. 1):

```
<LISTAI>
    <AI>
        <AINR>00</AINR>
        <TEXT>SSCC</TEXT>
        <TYPE>n</TYPE>
        <LENGTH>18</LENGTH>
    </AI>
    <AI>
        <AINR>11</AINR>
        <TEXT>PROD DATE</TEXT>
        <TYPE>n</TYPE>
        <LENGTH>6</LENGTH>
    </AI>
    <AI>
        <AINR>13</AINR>
        <TEXT>PACK DATE</TEXT>
        <TYP>n</TYP>
        <LENGTH>6</LENGTH>
    </AI>
</LISTAI>
```

As shown in Fig. 1, the application starts with a menu that offers the possibility to change configuration options, to choose an identifier scheme and the belonging items as well as the possibility to read and write transponder. Data@RFID supports the tag types ISO 18006-B and EPC Class 1 Generation 2. At the time of the project start, ISO 18006-B offered more capacity of the user memory, thus more data could be stored. After choosing an identifier scheme, the information stored in the XML file is used to build a graphical user interface with all contained identifiers. Fig. 1 shows the example of an activated Application Identifier scheme. After choosing the identifiers' Serial Shipping Container Code (SSCC), production date (PROD DATE) and pack date (PACK DATE), only the chosen identifiers are shown as part of the GUI. The information about data type and length of the identifier are used to prove the data before it is captured and stored to the transponder. The captured data is stored according to ISO 15434:

```
[)>RS05GS00123456789123456789GS11060708GS13070708RSEOT
```

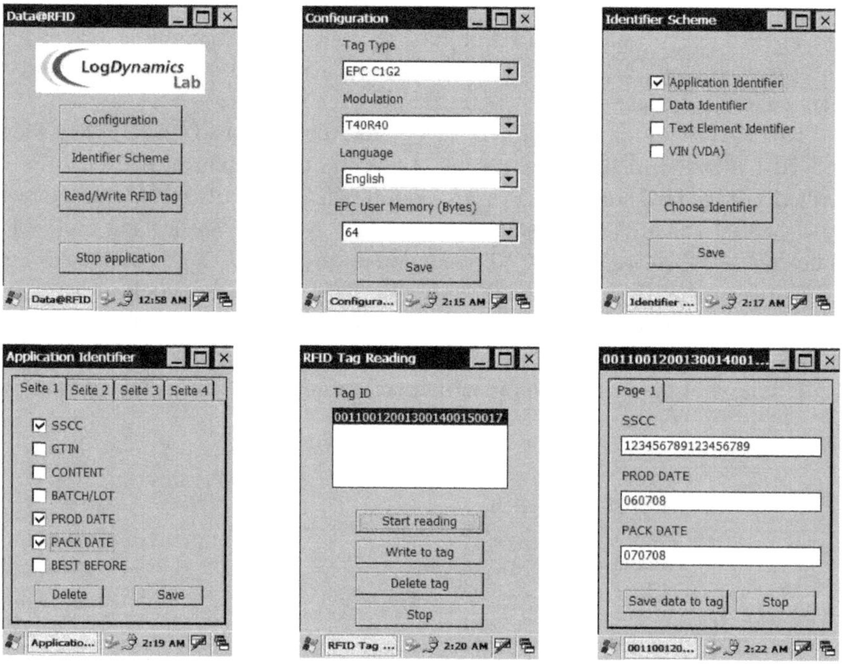

Fig. 1. Data@RFID – Application Identifier (Screenshots)

To prevent memory errors it is also necessary to know and validate the length of the user memory. The capacity can be chosen on the configuration page. The application can also read transponder without choosing an identifier scheme. In that case it compares if the data on the transponder corresponds to any of the provided schemes. If a specific scheme is identified, the data is displayed in the corresponding format.

The Vehicle Identification Number (VIN) is supported by Data@RFID as well. The application was extended and used to prove the concept of the VDA 5520 standard, which was also used as basis for a recommendation about the usage of RFID in

Table 1. Identifier Schemes

Identifier Scheme	Usage	Standard
Application Identifier (AI)	retail	ISO 15434, ISO 15418
Data Identifier (DI)	industry; courier, express and parcel	ISO 15434, ISO 15418
Text Element Identifier (TEI)	aviation	ISO 15434, ATA Spec 2000
Vehicle Identification Number (VIN)	automotive industry	ISO 15961, ISO 15962

vehicle distribution processes [4]. The VIN is stored in the EPC memory bank of the transponder. The EPC memory has to provide at least 144 bits to store a VIN according to VDA 5520. The VIN is 6-bit coded according to ISO 15692, while all other identifier schemes are stored using ASCII code.

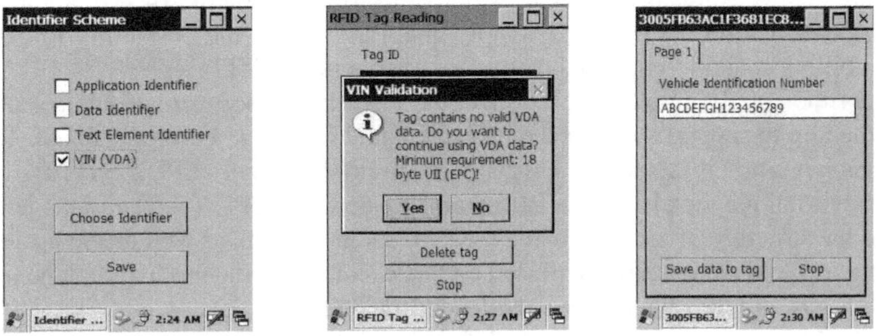

Fig. 2. Data@RFID – Vehicle Identification Number (Screenshot)

Fig. 2 shows some screenshots of Data@RFID showing the capturing of a VIN. Before capturing the VIN, there is a hint that the EPC memory has to offer a minimum capacity of 18 bytes. The VIN "ABCDEFGH123456789" (see Fig. 2) is stored on the EPC memory bank as follows [4]:

Table 2. Content of EPC Memory (Vehicle Identification Number, 144 Bit)

Word of EPC	Byte 1	Byte 2	Content
Word 0 (CRC)	CRC-Byte 1	CRC-Byte 2	without modification
Word 1 (PC bits)	49	03	length of EPC, AFI
Word 2	0A	4F	DSFID, precursor of VIN
Word 3	51	0D	rel. OID, length of VIN
Word 4	04	20	VIN, 6bit coded
Word 5	C4	14	VIN, 6bit coded
Word 6	61	C8	VIN, 6bit coded
Word 7	C7	2C	VIN, 6bit coded
Word 8	F4	D7	VIN, 6bit coded
Word 9	6D	F8	VIN, 6bit coded
Word 10	E0	00	VIN, 6bit - terminator

Within the LogDynamics Lab, Data@RFID is used to show that it is possible to combine different data schemes. The different items at the LogDynamics Lab, such as beverage and food pallets (retail), airplane parts, or car shipping labels, can be processed with one mobile application. Irrespective of the industry (e.g., automotive, aviation, retail) this single application is used to write industry specific data, which go beyond simple identification. When reading the corresponding tag data, the data scheme is automatically retrieved. This application shows that different standards can still be integrated through corresponding application development.

3 Retail Scenario – EPCglobal-Based Storage of Data-on-Network

This project was initiated in the LogDynamics Lab after a pilot project that evaluated performance increase and benefit compensation within supply chains using RFID [5]. Its goal was to evaluate the usage of the EPCglobal standards [6] for capturing and sharing data within the beverage supply chain as well as the possibility to extend it with a billing system to bill the costs of returnable items (RTI).

The EPCglobal standards are a collection of hardware, software and data standards with the aim to support structured data exchange throughout the supply chain. Therefore, each product is labeled with an Electronic Product Code (EPC). The EPC can be stored as a unique identifier on a RFID transponder. The EPC Tag Data Standards [6] define the structure of the EPC using the EPC memory bank as well as the usage of the free available user memory. Based on those standards, the products can be traced on their way and may be equipped with additional product data.

To build up a corresponding EPCglobal Network demonstrator we had to define:

- a part of the beverage supply chain to be represented;
- products and the corresponding types of EPCs;
- physical implementation of the scenario including locations and RFID readers;
- type and number of EPCIS events to be captured;
- logical implementation to capture, store and represent all EPCIS events.

A simple scenario containing bottler, wholesaler and retailer was built up in the LogDynamics Lab. It consists of four reading locations:

- Bottler: EndOfProduction;
- Bottler: OutgoingGoods;
- Distributor: IncomingGoods;
- Distributor: OutgoingGoods.

By means of this scenario, it is possible to follow the product after bottling until it is sold to the retailer as well as the way back. The data flow is represented in Fig. 3.

The sample product used with the demonstrator is a pallet filled with bottles provided by LOGIPACK for the pilot project mentioned above [5]. There are four dollies on the pallet, each of them filled up with six trays. Each tray contains six six-packs. The dollies are tagged by the company with a Global Returnable Asset Identifier (GRAI). Following the EPCglobal Tag Standards [6], corresponding GRAIs were used for the trays and a pallet was identified based on the Serial Shipping Container Code (SSCC). The test pallet holds 29 transponders to identify one pallet, four dollies and 24 trays. The bottles are not tagged.

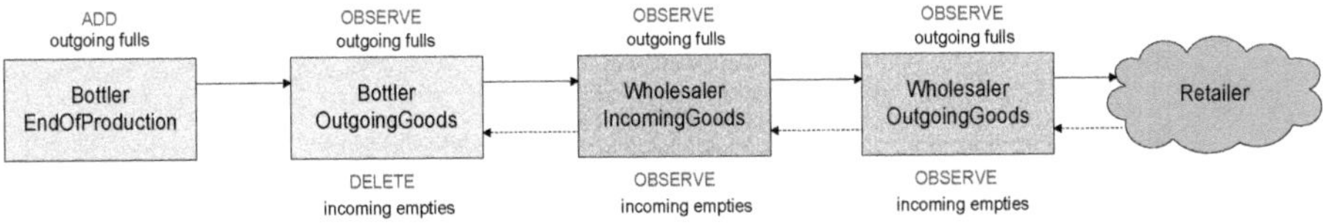

Fig. 3. Scenario of the beverage supply chain (participants, data flow, actions)

To illustrate this retail scenario a demonstrator was constructed. The physical buildup at the LogDynamics Lab contains four reader locations. The first station is realized as a pallet wrapper with an integrated RFID reader simulating the end of the production at the bottler. All other locations are realized with gates equipped with a reader. The test pallet can be moved with a forklift from the pallet wrapper through all gates and backwards to simulate the reverse logistics supply chain. Following the EPCglobal Architecture [6], all readers are connected to a RFID Middleware. For this purpose a Reva TAP device is used. It also offers a client to access the Application Level Events (ALE).

Within the EPCglobal Architecture, so called EPCIS events [6] are stored at each defined reader location. The EPCIS event contains the information about *What?, Where?, When?* and *Why?* of a product:

- *What* kind of product has been read? The EPCIS event captures the EPC that gives the information about the product type, for example GRAI or SSCC.
- *Where* was the product read? The EPCIS event stores the referring reader location.
- *When* was the product read? The EPCIS event captures a time stamp.
- *Why* was the product there? The EPCIS event stores the reason why the product was passing this location, for instance shipping.

There are four different types of EPCIS events:

- ObjectEvent;
- AggregationEvent;
- TransactionEvent;
- QuantityEvent.

The retail scenario uses only two of them. ObjectEvents are captured for the trays and dollies. AggregationEvents are defined for the pallets containing all belonging ObjectEvents.

As the EPCglobal standards and interfaces are only a recommendation, it is necessary to have an implementation of the EPCIS repository [6] to store all captured events. For this, the Fosstrak EPCIS [7] was used as it can be downloaded for free as an open source project. It is developed based on Java and MySQL and offers a repository following the EPCglobal standards and the possibility to implement suitable Capturing and Accessing Applications.

The complete system architecture is shown in Fig. 4. The Accessing Applications of the brewery and the retailer(s) are optional. The installations of Fosstrak EPCIS for the bottler and the wholesaler are done on virtual machines using VMWare Server. It is easy to extend the network with more participants using virtualization.

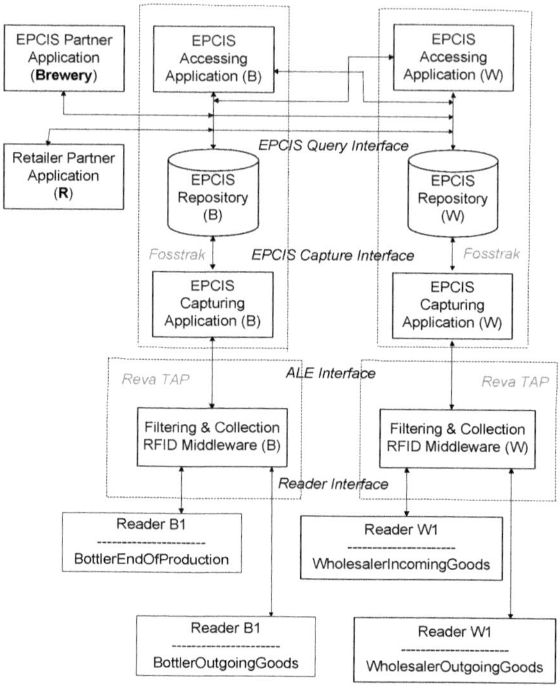

Fig. 4. System Architecture (EPC Network, based on [8])

There are two Capturing Applications for the bottler and the distributor. Both of them are designed to listen to different ports to get the ALE events provided by the Reva ALE client. The Capturing Applications transform these ALE events to EPCIS events that are stored to the EPCIS repositories (see Fig. 4).

Fig. 5. Product Data (Screenshots of Capturing Applications)

Additionally, the EPC Network demonstrator uses the extension mechanism of the EPCglobal specification to store product data. The capturing is included into the Capturing Applications of bottler and distributor (see Fig. 5).

Based on the developed demonstrator, research projects such as a billing integrated EPCglobal Network could be developed [8]. Additionally, the platform is used for educational purposes with corresponding lectures and student projects [9]. In those cases only small adaptions have been necessary to use the EPC Network demonstrator. For the usage with the integrated billing system the Capturing and Accessing Applications had to be extended to store and query data from the billing system as well. As there were only small test scenarios (up to three different pallets), no limitations of the performance has been recognized. Nevertheless, it seems to be useful to watch the performance of data processing in real-world scenarios with a huge amount of data, especially if it has to be transferred to existing Enterprise Information Systems [10].

4 RFID-Based Automotive Network

4.1 Project Description

The RFID-based Automotive Network (RAN) is a research project with the aim to control the processes of the automotive industry in a transparent and optimal way. RAN is funded by the German Ministry of Economics and Technology (BMWi) and composed of different partners involved in the production and logistic chain, for instance manufactures, suppliers, logistic companies, IT companies and research institutes. The project is organized as a matrix which consists of seven use cases and five work packages (see Fig. 6).

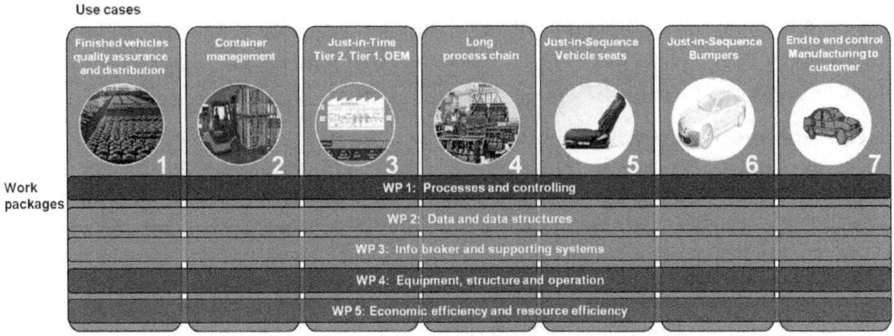

Fig. 6. RAN matrix – Use Cases and Work Packages [11]

Although the focus of this consortium is working on real-world use cases, there are three demonstration centers belonging to the RAN project as well. These demonstration centers have the objective to represent the RAN overall concept and to demonstrate its functionality under lab conditions. Therefore, methods like virtualization, pilot applications, demo buildups and learning games are used to implement these demonstration centers. One of them is situated at the BIBA that is involved into the project as a research institute. Its main tasks are:

- representing the overall concept of RAN;
- visualizing the processes of Use Case 1 (Finished vehicles);
- demonstrating the wearable computing system "easyTracing System" (based on [2]);
- establishing a learning lab.

This paper focuses on the learning lab which will be described in the next subsection.

4.2 Learning Lab

The RAN Learning Lab targets new participants of the project, for instance distributors, suppliers or manufacturers. It provides information about the methods and processes of RAN and explains the used technologies, for example RFID or sensors. The data exchange in RAN is based on the EPCglobal architecture. It is implemented as an Infobroker using the EPCglobal specifications and extending them by the requirements of the RAN processes. Therefore, the learning lab has to explain the EPCglobal architecture and its usage in RAN. To ease the integration of new partners, the learning lab offers a learning game to interested participants. It can be played on mobile tablets making the complete network with its processes come alive. The game can be played in two variants:

Virtual Automotive Supply Chain: This is a serious mobile 3D game that allows a user to move a car throughout the supply chain based on the RAN processes. In this way the EPCglobal architecture will be introduced and virtual EPCIS events will be captured. The game can be played with up to six persons. The supply chain can be configured based on the needs of the player. The configuration is based on the RFID roles: manufacturer, logistic company and distributor. The whole game is visualized on a 52" monitor equipped with a computer containing the belonging EPCIS repositories of the participants. This variant of the game focuses on demonstrating the processes and methods of RAN.

Real Automotive Supply Chain: This variant extends the first version of the learning game. The mobile game on the tablets will be connected to a real world scenario. The "real world" consists of RFID readers and gates as well as remote-controlled cars. The cars are labeled with RFID transponders using the electronic Vehicle Identification Number (eVIN) that was defined in the RAN project with the aim to store all VIN-related information (according to ISO 3779):

- World Manufacturer Identifier;
- Vehicle Descriptor Section;
- Vehicle Identifier Section.

While this aim is similar to the VIN implementation in the Data@RFID project, there is a difference between them: The definition of the eVIN is based on the EPCglobal Tag Data Standards [6], while usage of the VIN in the Data@RFID project was based on ISO 15961.

During the game, EPCIS events will be generated in real-time at the reader locations and captured into EPCIS repositories. This variant of the game is focused on introducing the RFID technology. Its configuration is less flexible as the game has to match the real-world scenario.

The option to configure the supply chain depending on the role of the participant should help new participants to realize the whole network from their own point of view. Therefore, it allows an easier understanding of company interaction and information exchange. The integration of real RFID hardware and transponders offers the opportunity to get directly in touch with the involved technology.

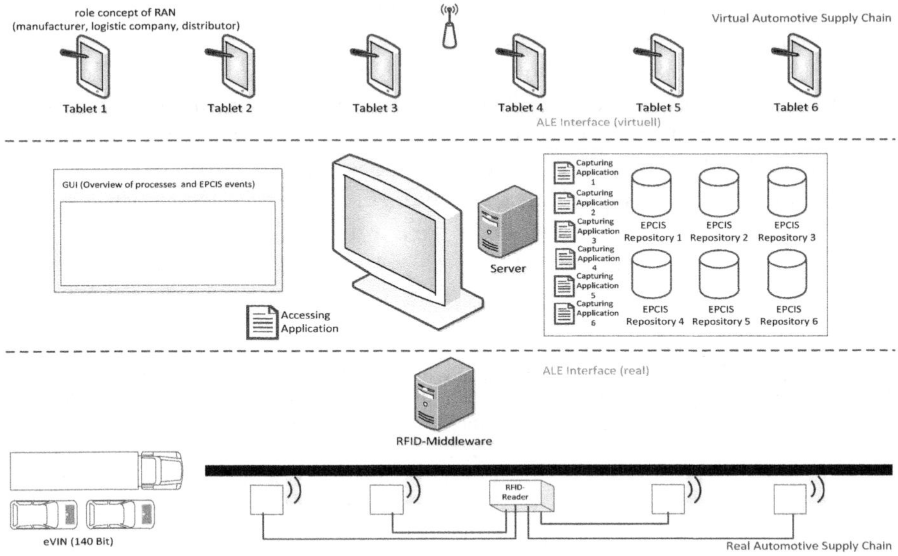

Fig. 7. System architecture (RAN Learning Lab)

Within the RAN project the Learning Lab helps to explain the overall concept of the project and to visualize the distributed processes throughout the automotive supply chain. While the various use cases of the projects are working on real-world scenarios, the three demonstration centers are focused on demonstrating and teaching the defined processes, data exchange concepts and the usage of new technologies.

5 Conclusion

The paper introduced three lab projects performed in the last years. All of them researched the structured data exchange based on RFID but were targeting different aims. Comparing their objectives, methods, results and the impact of the lab environment (see Table 3) we can concern the following scopes:

- testing the opportunities of new technologies;
- validating, extending and introducing standards;
- developing interfaces to established systems (e.g., billing software);
- demonstrating and teaching new technologies and standards using simulation models and (small) scenarios close to reality.

The history of the introduced projects shows the changing impact of a lab environment. Although all of them target the same topic, the objectives are changing. At the beginning the RFID technology is rather new and the data is less standardized. Therefore, the first project researches the usability of different standards to support data exchange within the supply chain. In the last years the EPCglobal standard has gained considerable acceptance in research and industry. As data-on-tag storage in the first project seems to be difficult due to technical limitations, the next project focuses on validating the concept of the EPCglobal standards that follows the data-on-network concept. The proof of concept was successfully established. The standards were further developed and can be used under real conditions now. Therefore, the third project steps from implementing pilots to teaching users in a lab environment. This example may show that a lab environment has to be flexible due to changing conditions.

Table 3. Comparison of lab projects

Project	Objectives	Methods	Results	Impact of lab
Data@RFID (2007)	data-on-tag storage, providing data with a product	evaluation of common ISO standards (logistics)	pilot application, problem: technical limitations due to less user memory	proof of concept, identifying limitations, usage of pilot to prove the concept of VDA 5520 specification during development
EPC Network Demonstrator (2008)	data-on-network storage, providing product data within a network, supporting RTIs (beverage supply chain)	evaluation of EPCglobal standards, virtualization	extendable demo architecture, testing facility for new developments (e.g., integration of a billing solution), platform for teaching	proof of concept, defining requirements of a billing interface, usage of demonstrator within other lab projects
RAN Learning Lab (2010)	explaining the RAN overall concept including processes, data exchange and technologies (automotive supply chain)	virtualization of processes and data flow (EPCglobal-based), real world scenario (RFID)	serious mobile learning game	providing a learning environment adaptable to the special requirements of new participants

Acknowledgement. This work was funded by the Federal Ministry for Economics and Technology (BMWi) under the reference number 01MA10009 "RAN – RFID based Automotive Network".

References

1. Lodewijks, G., Veeke, H.P.M., López De La Cruz, A.M.: Reliability of RFID in Logistic Systems. In: IEEE International Conference on Service Operations and Logistics, and Informatics, pp. 971–976 (2006)
2. Böse, F., Piotrowski, J., Scholz-Reiter, B.: Autonomously controlled storage management in vehicle logistics – applications of RFID and mobile computing systems. Int. J. RF Technol.: Res. Appl. 1(1), 57–76 (2009)
3. Rezwan, A.: Reverse Logistics: RFID the key to optimality. J. Ind. Eng. Manag. 4(2), 281–300 (2011)
4. ODETTE: Automotive Supply Chain Best Practice Recommendation: RFID in Vehicle Distribution Processes (2010)
5. Uckelmann, D., Hamann, T., Zschintsch, M.: Performance increase and benefit compensation in supply chains by partial information sharing and billing based on identification of returnable transport icons. Int. J. RF Technol.: Res. Appl. 1(1), 23–43 (2009)
6. EPCglobal standards, `http://www.gs1.org/gsmp/kc/epcglobal`
7. Fosstrak EPCIS, `http://www.fosstrak.org/epcis/index.html`
8. Uckelmann, D.: A billing integrated EPCglobal Network. In: Internet of Things 2010 Conference (2010)
9. VLL – Das Virtual Logistics Lab, `http://www.virtual-logistics-lab.de`
10. Su, X., Chu, C., Prabhu, B.S., Gadh, R.: Creating a RFID Data Integration Framework for Enterprise Information System. Intern. J. Internet. Protoc. Technol. 4(4), 221–231 (2009)
11. RAN-Infoflyer, `http://www.autoran.de/fileadmin/autoran.de/data/Dokumente/RAN-Infoflyer.pdf`

How AutoID Processes Shape the Internet of Things: The OpenID-Center Development Process

Benedikt Mättig, Martin Fiedler, Alexander Hille, and Björn Anderseck

Fraunhofer Institute for Material Flow and Logistics IML
Joseph-von-Fraunhofer-Str. 2-4
44227 Dortmund, Germany
{benedikt.maettig,martin.fiedler,alexander.hille,
bjoern.anderseck}@iml.fraunhofer.de

Abstract. The openID-center as an open integration platform for intra logistic systems focuses its development process on applications and processes. Different market-ready solutions like Barcode, HF-/UHF RFID, RTLS and telematics may be evaluated by industry customers within a test-bed with different facilities. Current research results are brought in the development process to give an outlook to future solutions. Customer feedback on future solutions is given as an input to the corresponding research project.

Keywords: Internet of Things, OpenID-Center, Process Design, Technical Feasibility, Piloting, RFID.

1 Introduction

The openID-center is an open integration platform for supply chain technologies and especially AutoID. Integrated in the Fraunhofer Institute for Material Flow and Logistics IML in Dortmund, the openID-center offers a 1.500m² testing area for 'state of the art' logistics solutions. The research activities are mostly application focused and deal with 1D/2D-Barcodes, HF/UHF RFID and RTLS technologies as well as software developments. Every single project has one common aim: shaping the Internet of Things. This shaping is done by a predefined development process that is used in almost every project and takes place mainly within the lab. This process consists of three main modules: Process-design, technical feasibility studies and piloting.

The process design marks the beginning of every project. After analyzing customers' current processes, an integrated AutoID process is designed. The process describes the material flow on the one hand and the information flow on the other hand. Based on these data, a cost-benefit analysis follows up. This analysis is split up into three parts. The first analysis calculates the costs and benefits which have direct impact on the customer. However, especially in the case of RTLS, for example, the long-term impact on processes is hidden. In the openID-center development, these

Dieter Uckelmann et al. (Eds.): ImViReLL 2012, CCIS 282, pp. 134–147, 2012.

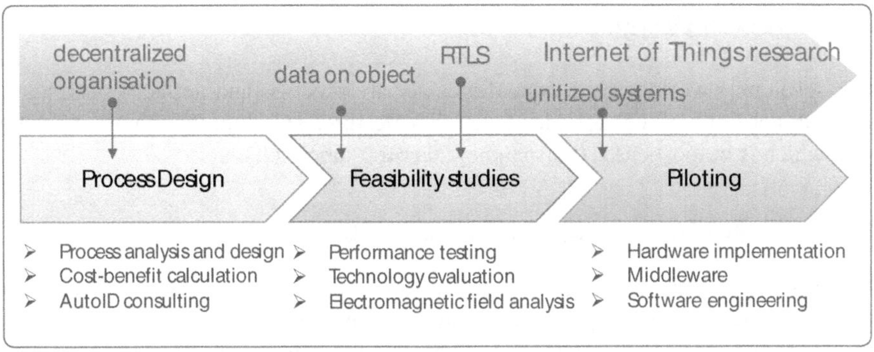

Fig. 1. Overview of the openID-center development process

hidden and long-term benefits become visible for the customer by game changing processes, which enable decentralized logistics with the Internet of Things. The second module is the technical feasibility study, where the openID-center emphasizes its strength. The possibility to test several RFID-based identification- and localization-technologies in an industrial-related environment proves the quality of concept and simplifies the implementation. All current available hardware devices in the lab are implemented via the openID-middleware. That allows high-class and fast tests with the largest bandwidth of hardware. The third module is piloting. Implementing the developed process with the tested hardware at customer facilities is the proof of concept for the openID-center development process and for the Internet of Things. The exact characteristics of the three modules and how they shape the Internet of Things will be described in the following chapters.

Other Internet of Things technology lab approaches are comparable to the introduced one. For example, the recommended procedure for implementing RFID or related technologies at the RFID center of application in Munich is also separated in three main steps [14]:

- The first part "analyses of potentials and processes" is quite similar to the openID-center main module process design, which implies the cost-benefit calculation as potential indicator.
- The second part "feasibility studies and technology selection" complies more or less with the openID-center modules feasibility studies and piloting. The approaches, the used facilities, software systems and possible simulations are different and for both labs unique.
- "Attendance of RFID projects" is the third part and implies also steps like testing and monitoring as the openID-center piloting module. Furthermore, preparing of requirement documents and valuation of proposals from technology integration companies is described.

2 Process Design

The developers working at the openID-center are process- and application-designers. They do not think of technologies when they design a solution, they think of the final aim, which is in most cases transparency over the supply chain.

2.1 Introduction to the Internet of Things Vision

The Internet of Things became a research topic of high interest lately, so numerous industrial and research organizations are working on the different aspects of the vision. Depending on the specific project, the outcome of the work is focused on theoretical models or practical applications. This is why there exist different views on what the Internet of Things is and several approaches on how to realize possible applications.

Generally, the vision consists of combining the physical world of objects (a product, a transport item or cattle) with the virtual world of the Internet. Users may work with the physical objects by using a virtual representation which is connected to the Internet. The other way around, each physical object may connect to everything which is attached to the Internet, be it another physical object, a software daemon or a human user working on a web browser. In a simple way, only data may be attached to physical objects in a direct or indirect way. Physical objects become real "smart objects" in the way they are able to compute their own actions (e.g., use of some resource) on locally sensed values and communicate with others [1].

With the advances in miniaturizing of hardware, it is only a matter of time for Mark Weiser's vision of *Ubiquitous Computing* [3] to become a reality. Microcontrollers, sensors, actuators and communication devices get smaller and more powerful while using less energy. Today, passive RF transponders are used as a rewritable data store which may be attached to different everyday objects. RFID transponders may compute simple tasks by themselves, for instance read attached sensor values and store them.

Besides work to be done on hardware development for smart objects, the integration of these in business processes is another part of active research. There are standards for RFID integration (e.g., EPCGlobal), while standards for smart objects in the context of the Internet of Things are often missing.

The logistics and transportation domain is a valuable application scenario for further Internet of Things developments. One of the main problems in logistics is the synchronization of material and information flow of objects. A future goal is to combine both flows by smart logistic objects (e.g., items, boxes, pallets) to maximize supply chain transparency [2].

Within research at openID-center, different Internet of Things demonstrators are implemented to provide real world feedback of concepts. For example, a modular built conveying system that is controlled by distributed software agents which represent conveyors, RFID sensors, boxes, etc. follow a specific mission to build a control system in total. A possibility to calculate the near real-time material flow out of sensor events was shown in [4]. The research effort on software agent-based material flow systems resulted in a product which is commercially available today[1].

Current research projects which are being worked on in the openID-center are mainly the following: The *Internet of Things Architecture (IoT-A)* project [7] is the European Lighthouse Integrated Project, which is working on the formal definition of an architectural reference model including an initial set of key building blocks. Parts of the projects use case demonstrations in two different domains will be implemented within the openID-center. Within the *smart reusable transport items (smaRTI)* project [8], hybrid AutoID processes are developed. In contrast, *Dynamic Container Networks (DyCoNet)* [9] focuses on hardware development of smart ULD containers, which are used in airfreight. A GSM/GPS and low power short range radio-based smart object allows for near real-time tracking of ULDs and its contained goods.

Research results of these projects are quite early put into practical evaluation with the use of the openID-center test bed. In this way, a useful input to (future) possibilities of Internet of Things technologies is given to industry customers and feedback on practicability is given to the future work on the research project. It is often a direct interchange of information and experience between researchers and end-users from the industries to avoid developments which are not useable from a practical point of view.

2.2 Process Analysis and Development

Every process that is developed at the openID-center starts with the analysis of the customer needs. The most important thing is to find out the aim a customer has, not the technology he thought of that should bring the solution. Beginning with that basic requirement the characteristics of the Internet of Things influence the process design.

The customers' current processes are modeled in needed notations (e.g., BPMN, flow chart) where the modeling depth may vary upon the needed problem. Formally, the material flow with its corresponding information flow is put into one model. In this way it is possible to find an integrated process which allows the use of different technologies. Depending on the process environment and needed integration depth, the following technologies are considered:

- Barcode, 2D-Code
- passive RFID
- passive/active RFID with sensors
- sensor/actuator networks
- GSM/GPS-based telematics

Each technology allows for different integration of material- and information flow, as a result different process models for different integrated technologies are built.

Future systems based on sensor nodes or telematics systems allow the integration of autonomous processes. These smart objects follow and monitor a sub-process by themselves and interact with other process objects. As an example, the DyCoNet

---

[1] `http://www.thingtelligence.de`

smartULD container [9] will run such sub-processes. There is no need for a central system controlling all of the ULDs, instead the smart ULDs fulfill subtasks autonomously. These kinds of processes are coming more and more into the minds of the customers.

In a final step, the cost-benefit analysis follows up, which gives the customer a detailed insight on short-, as well as long-term results.

3 Technical Feasibility Studies

Technical feasibility studies are one major research part in the openID-center. The main tested technologies on an area of 1.500 m^2 are RFID systems (readers, antennas and tags), different types of barcode scanners (typical barcode, 2D-codes and direct part marking technology) and localization (ultra wideband). The lab provides a platform for manufacture and reseller independent testing. In case of RFID, for example, it includes more than ten different reader systems, a high number of standardized tags on the market, individual tag developments in cooperation with partners and self-developed software environments. The technology focus lies on UHF RFID. In the following part of this paper, the most important test facilities and those with the highest impact on the future application design are described in more detail.

3.1 OpenID-Center as Environment for RFID Performance Tests

In order to test the technical usability of RFID solutions in certain use cases, the openID-center provides an environment for performing these tests.

This environment consists of a straight automatic conveyor track, which is 15 meters long and can carry up to 1.5 tons of goods on a movable platform at a maximum speed of 4.5 m/sec. It can be equipped with a large variety of RFID readers and antennas for creating several different test scenarios. The entire hardware is connected to an IT-system which controls the whole test.

The controlling IT-system is represented by the openID-testsuite. This software, which was developed for controlling and managing the performance tests, offers a whole project and test management user interface, which can be used for planning, processing and analyzing tests. Test results are presented in an easily understandable 3D-view and can be exported in several formats for later processing. While executing a test, the testsuite controls the conveyor and the used RFID reader. Because there are various readers from different manufacturers, which all have their own interior software and protocols, a specific openID-middleware was developed, which offers an unified interface for every implemented reader and serves as a general translator and coordinator between any kind of reader and the testsuite. For every test, the testsuite allows the tester to change signal strength, amount of active antennas and speed of the conveyor as well as many other parameters. The target of these tests is to verify the most suitable location and combination of reader, antenna and tag for a specific scenario.

Two different test scenarios are performed on this system: On the one hand bulk reading tests which simulate common gate use cases and on the other hand distance tests which will give a general overview about the maximum reading distance of a selected reader tag combination.

The most suitable location to have an RFID tag put on is different for every product, load carrier, signal and tag type. Specific properties of a product may interfere with the antennas' electromagnetic field, complicating communication between antennas and transponder, and often RFID tags have to be aligned to the antennas' signal in a certain way to be able to reply to it. The aim of a bulk reading test is to find the "sweetspot" on a product that will preferably eradicate these interferences. The pallet is loaded with a packing unit of tagged products that is typical in size and structure of stacking for the logistic tasks it would go through in reality. Now, the pallet moves across the test track and when passing the antennas, the RFID tags, activated by the electromagnetic field, will send out a reply. The signals received by the antennas are sent to the reader, which interprets them and forwards this interpretation to the computer. The testsuite now provides the operator with an analysis of the test, consisting of the tags' replies. If there is a reply missing, this either means the antennas' signal was not able to reach the tag or the reader misinterpreted messages sent to it by the antennas. The software will show that only a certain percentage of the actual amount of goods has been recognized. The specific tags which did not answer, can be explicitly identified and even be located in a 3D-animation of the pallet's load to locate problematic positions of tags inside the load. Parameters can now be changed to achieve better results in future test runs. Wanted results consist of 100% recognition of tags at a preferably great velocity of the pallet, while the energy needed to activate all tags is kept comparatively low. After the results have been evaluated, a combination of tag type, signal strength and pallet speed that scored the best result is recorded as a recommendation to the customer.

Fig. 2. Test setup for bulk testing

Different RFID tags operate on different signal strengths and within different distances to the antennas. To find out how these requirements vary in relation to the tag's location and specific features of the product IML runs distance measuring tests. For these tests, a tagged product unit is put on the test track's platform. Now, the single antenna at the start of the track sends a signal across the track while the platform moves away from it. While the signal still reaches the tag, it provides the tag with the energy necessary to send an answer back to it. As soon as the tag does not reply anymore, the distance between the platform and the antenna is recorded as the maximum distance. These tests are done in both ways: starting from the antenna moving away from it as well as from the end of the track going towards the antenna. This is important, since an RFID tag needs a certain amount of activation energy to initially send an answer and this level of energy is only reached after the tag has been within the antennas electromagnetic field for a certain amount of time. Usually, these tests deliver two kinds of maximum distances for each RFID tag: one for maximum and one for minimum signal strength. Depending on the shape of the tagged object, several transponders can be tested at once.

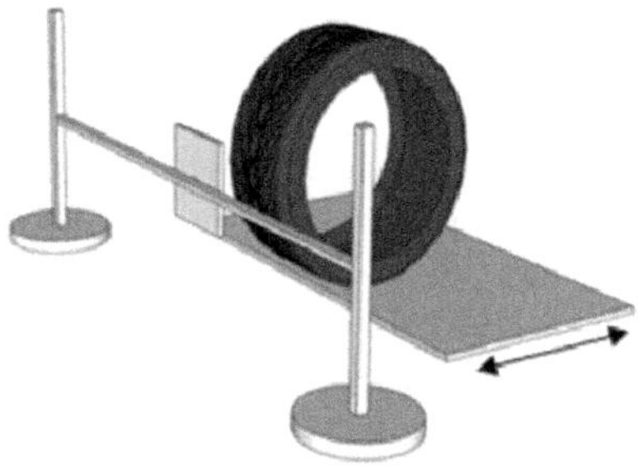

Fig. 3. Test setup for distance measuring

The different test scenarios are reflected in several problems, which are often submitted by the industry. For example, one project was devoted to the identification of pallets with bottles filled with water. Many other examples and references can be found here [15]. This identification should be implemented with the help of RFID. The difficulty of this challenge was that water has a very high permittivity so the antennas' signal was likely to be disturbed by the content of the bottles. Therefore, in extensive tests an optimal position for mounting the tag on the pallet had to be found to realize the requirements.

In another test the challenge was to tag beer kegs with RFID for reading them in bulk while the forklift carries them through a gate. The problem here was that the beer kegs completely was made of metal which can reflect or shield the antennas' electromagnetic field and thus rendering them unable to reach and activate RFID tags. A second requirement was that the tag had to be mounted direct on the surface of the kegs because the tag needed to be properly protected from physical damage. Because of that, many different tags and modes of fixing had been tested. To simulate a forklift, the platform was set to drive through the antenna field with a typical forklift speed.

3.2 Active Ultra Wideband Location Technology

Ubisense provides an active real-time location system for indoor purposes in harsh environments. It is designed for process and asset monitoring, visibility, tracking and control in different application scenarios like automotive, aerospace, transport and military [10]. The system is also able to create three dimensional location data in a high accuracy. It consists of three parts: the receiving antennas, which are distributed over the location space, the active transponders sending the signal in free space, and the backend server for configuration and control. For the precise locating, Ubisense combines Angle-of-Arrival and Time-Difference-of-Arrival technology, which makes it possible to locate a tag within 15 cm in three dimensions [5]. The antenna detects UWB pulses from the transponders. All Sensors are connected via Ethernet to the server and also to each other [6].

The system was installed in the openID-center in 2007. Today it covers nearly the whole space of 1.500 m^2, with special setups for single applications in the hall. Overall, 14 antennas are installed. Meanwhile, the setup was redesigned and rebuilt for different kind of purposes:

- tracking of persons in wide space,
- vehicle (especially forklift) tracking,
- returnable asset tracking (boxes), and
- commissioning control.

Each use case defines special requirements for the RTLS relating to accuracy, fast response messages and the insensitivity to harsh environments like metals. The main goal was to show the general possibilities of the technology, to prove the technical feasibility and to show needs for further developments of the system in these applications. The tracking of returnable assets is a good example for the achievable high resolution. Single boxes are located on a typical conveyor belt by one active transponder equipped on each box. The system achieved a resolution of approximately one conveyor lane, which is 30 to 40 cm wide, even in this environment where mostly all parts of the belt are made out of metal.

In addition, the system is tested for forklift tracking, which has a big impact on practical logistic processes. The position information of the forklift can be used for optimal driver guidance through the warehouse and for automatic booking processes, when goods are identified by AutoID systems and linked to the forklift. With additional sensors for detecting if the goods are on or off loaded, manual scans or submissions can be saved.

Another demonstration and possible use case for future developments is a commissioning control. The system locates where the worker picks in the shelf and compares this information with a known commissioning order and the shelf organization. In combination with a pick-by-voice system, a complete primary data entry without complex human-machine interactions is possible. This application was developed in cooperation with the engineering team of Ubisense and shows potentials for the future.

3.3 Testing of New Forms of Passive UHF RFID Systems

In the middle of 2011, one of the biggest test environments for the Mojix RFID system was build up in the openID-center. Mojix works also on passive UHF Gen2 tags like classic UHF readers. The system is especially designed for the detection of signals in difficult environments, where the signal to noise ratio is very low because of reflecting or absorbing materials in the environment of an RFID system [12]. It uses innovations in iterative signal processing, space time array techniques, smart antennas, digital beam forming, and digital packet radio breakthroughs for achieving a 100 times receiver sensitivity of today's UHF passive solutions [11]. One of the big differences between Mojix and classic interrogators is the system architecture. Conventional RFID systems integrate, transmit and receive functions in a single reader device. Mojix concept disassociates the transmitting and receiving part of the RFID system. Combined with a robust mechanism for the detection of weak signals with severe phase and timing ambiguities, this property can be achieved.

One, so called, Mojix STAR receiver covers nearly the complete space of 1.500 m^2 in the openID-center. For the testing purpose, a space of approximately 150 m^2 can be equipped with transmitting antennas in a height of 5 m. A rack out of 3 point traverses and special antenna brackets provides a high flexibility in testing different kind of system designs for dock door detections and locating purposes of passive UHF Gen 2 tags.

With this testing facility, the main questions in context of the practical use of the Mojix system can be answered:

- What is needed for setting up a complete Mojix RFID interrogator system in a warehouse?
- Which transponders can be read with a high performance in different environments and on different kind of objects (especially returnable transport items)?
- How accurate is the system in locating passive UHF transponders in different antenna layouts, on different transport items and in different process scenarios?
- What kind of impact has the environment and the process steps on the performance of the system?
- How can the system be tuned through adoptions of the logistic process?
- What kind of data interface is provided by the system and how can this be integrated into different information systems?

All of these points listed above must be seen from a different view depending on the underlying use case. Chapter 4.2 describes how the system was tested and further developed in context of a use case in the air freight distribution center of Lufthansa Cargo in Frankfurt.

The Mojix system is integrated in our existing test software environment, the openID-center middleware. From that point, the information provided by Mojix can be used individual through a .NET API. This enables the development of different test software systems for individual purposes, for example calculation of the location resolution or read rate for dock door identification scenarios. A visual analysis tool allows the graphical presentation of location events in a virtual 3D model of the openID-center in real time.

Fig. 4. Simple localization test setup at openID-center with 100 tagged boxes

So far, the achieved results for locating of passive UHF RFID tags with Mojix are different and strongly dependent on the scenario properties. For example, with a tag on a plastic box, placed within the test area, a resolution of less than 1 m can be achieved in most cases. In other tag orientations, for example on the side of the box in a vertical alignment, the readability decreases and therewith also the resolution. The tests have also shown that the system is still restricted by the typical influences of an RFID system through reflections and absorptions, more on the sending part then on the receiving one. Also, with metal obstruction like forklifts or racks between the tags and the receiver, which leads to none line of sight between them, reads are possible. The main difficulty is the energy supply of the tag by the transmitting antennas. Through reflections, tags get powered which should not be powered during localization and vice versa. This makes the usability of the system highly depending on the individual use case and feasibility studies like that one for Lufthansa still necessary.

4 Piloting

The openID-center as part of an applied research institute accompanies its customers up to a pilot implementation. Complete technology roll-outs are done by system integration partners with consultancy of Fraunhofer IML. Especially the own software development with the openID-center Middleware enables the team to implement pilots very fast and easy. Existing interfaces to IT-Systems like warehouse management or ERPs improve the showcase. The following examples give a deep insight into two projects of the openID-center.

4.1 Piloting Example: SKF

One success story which demonstrates the impact of the openID-center on logistic processes in a good way is a pilot project for SKF. SKF is a ball bearing producing company with single producing and central distribution locations worldwide. The project was concentrated on two sites, one in Germany (Schweinfurt) and one in Austria (Steyr), to setup a best practice show case for the use of RFID within the SKF group. Produced ball bearings of different types are stored in special carriers, so called GSPs (Group Standard Pallets). These GSPs have a size of a quarter of a euro palette and are available in three different heights. After packaging, each GSP was equipped with one label which included five different barcodes: the unique box number, the weight, a batch number, the date of production and the origin-code (e.g., Steyr). The GSPs are loaded on and unloaded from a truck. Because of a missing link between the IT-systems of the two locations, each barcode of a GSP must be scanned while unloading the truck. This is a major effort and happens often after and not simultaneously to the physical unloading process.

The main idea was to substitute the manual scanning process by automatic identification and data transfer via RFID. For that issue, a high read rate for dock door processes, a bulk identification of up to 8 GSPs, which could be transported at the same time, and an untypical data storage strategy are necessary. The process analysis and the resulting profile of requirements leads to implement UHF RFID technology with a storage capacity of 96 bit EPC-memory and 512 bit user memory. To that time and until today, the use of these Gen2 tags in similar applications is not known. The future process design, which was developed together with SKF, includes an initializing of the tag on the GSP after the packaging process and identification points at goods out, goods in and automatic high belt storage locations.

First task of the project was to find out if this target could be achieved from a technical point of view. This means to answer the questions:

- How fast can the full tag memory be read in a transition?
- How reliably is the bulk identification with different amounts of GSPs?
- What is the best performing RFID system (interrogator and tag)?
- What is the best mounting position for the tag?
- Is there an influence by different kind of ball bearings which are stored in a GSP?

The test facilities within the openID-center provide a very good test platform for this kind of questioning. Most of the tests were done with the equipment described in chapter 3.1.

After the technical feasibility was proven in the IML lab, a big field try followed where interrogators at dock doors and packaging stations where installed and custom software tools where developed. The transponder initialization happens at the packaging station. When the GSP is in the allocated position, the employee attached a tag on the GSP. The employee starts the writing process with a smart software tool on a computer terminal at the workplace. Successful and not successful writings are reported and locked. In addition, this and every other process step is stored in a central database at IML over internet. From this point on, the complete box data can be read automatically by RFID in the supply chain:

- loading on the truck after packaging,
- unloading from the truck in the warehouse, and
- storing of GSPs in an automatic high belt system.

This new logistic process design and special use of UHF RFID technology, achieved faster lead times, a higher data quality and accuracy in real time, a reduction of data capturing errors and a high acceptance by the employees. The next step is to enhance this project on other locations of SKF and also to include external suppliers.

4.2 Testing Example: Lufthansa Cargo ULD Build-Up Use Case

Air freight is the most important business segment of Lufthansa. Lufthansa Cargo is one of the leading providers for global transport services. Approximately 1.81 million tons of freight and post were transported in 2010. Lufthansa operates three cargo hubs in Europe. The biggest and most important one is located in Frankfurt am Main [13]. From a logistics perspective, the daily business comprises the handling of single deliveries on a euro-pallet or piece level during a reception process, the consolidation of single shipments on loading devices like air freight pallets or containers (build-up), the loading of airplanes and the whole process in reverse order. The potential for mishandlings or wrong documentation in these processes is very high. Barcode labels are attached on every shipment which is received and also on every loading device which includes several shipments. Every storing and/or loading process of goods should be documented and posted in the backend system by scanning. This leads to increased process times and is no guaranty for correct and high quality documentation. The barcode solution provides potential for misuses. For example, all packages which should be loaded on the pallet are scanned before they are physically really loaded and after scanning they get loaded on a wrong air freight pallet or are forgotten. In the end, the carrier as the owner of the shipments and the client of Lufthansa cannot be informed in real time about the actual shipment status of his goods. In worst case, shipments get loaded on the wrong pallet and/or miss to be loaded on a flight. Because of this, Lufthansa thought about a technology which makes it possible to track each single piece of a shipment in the warehouse without manual identification processes through handhelds or any other devices operated by the ground handler.

On the one hand, the technical solution should be relatively cheap in terms of transponder costs and on the other hand guarantee a locating of goods in the warehouse on floor storage locations, in a rack and during the build-up process on defined zones to link airfreight pallets and single shipments in the backend system. For that purpose, a localization system like Mojix seems to be the right technology. The main task for the openID-center team was to find out in general, weather Mojix is the right technology from a technical point of view to achieve these goals and how can the system be tuned for an optimized functionality.

In the first test stage, the general resolution of the system under optimal circumstances on different positions within the rack was analyzed. Therefore, a recommended test setup from Mojix was used, where overall 9 x 4 antennas were installed with a distance between the center of each antenna chandelier of approximately five meters in every direction. The antennas were mounted in a height of approximately five meters with a pitch of 45 degrees.

The test procedure included over 100 single static[2] measurements for each position. The achieved results were strongly depending on the test scenario and the system design. Fig. 5 shows the results for different tag positions.

[2] The tag was positioned and approximately 30 seconds thereafter the testing started.

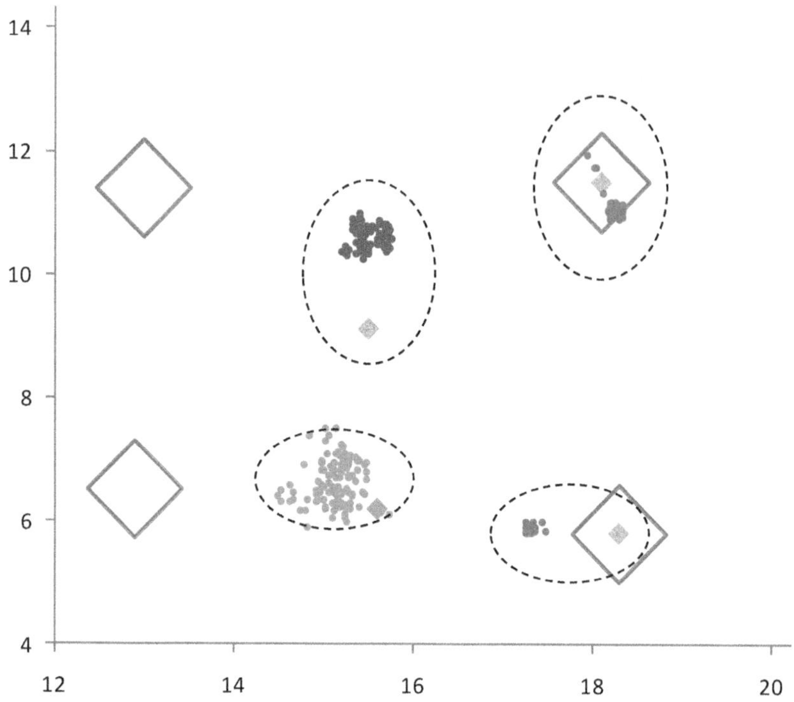

Fig. 5. Visualization of location dates from Mojix. The rues represent the mounted chandeliers described above.

For all test scenarios, accuracy and standard deviation were calculated. The influence of typical RFID disturbing objects in the near surrounding of the tags like forklift and airfreight containers was not that significant. In contrast, the parallel use of other RFID devices like handhelds caused a wide spreading of the single measurements and, as a consequence, is a big factor of disturbance.

In further tests, the influence of tag alignment on different types of objects with different content (metal and liquid), different types of standard commercial available Gen2 tags and different amounts of tags in the surrounding were analyzed. The achieved results lead to new antenna setups and further developments of the underlying software algorithms to optimize the system for the process with the highest requirements on the technology – the dynamic build-up process. Traditional locating configurations of the Mojix system obtain their limits in this highly flexible scenario where pieces in every size and property are packed individually and often as fast as possible on the airfreight pallet by a ground handler.

Finally, the technical feasibility in this application scenario was successfully analyzed and lead to further developments of the technology. The use of Mojix instead of the described barcode systems will change the future handling processes in the hubs and will have a direct impact on the logistic organization of Lufthansa. Documentation and system bookings get more and more automatic and high qualitative. Also, new strategies of working could be possible, based on the new transparency. In the near future, the current warehouse should be equipped partly with the technology to prove the concept and gain first experience in the different use cases. From the view of Lufthansa, the potential for a roll out on future hubs and other locations in the coming years is big.

5 Conclusions

The openID-center development process from process design to feasibility studies and piloting is improved by direct influence from research. All research activities are geared to the Internet of Things characteristics. Results from that research, such as methods, processes, applications or technologies have direct impact on customer projects. Future research topics evolve out of the customers' needs. Therefore, the openID-center integration platform is of high value for technology partners as well as application partners. The scheme describes the parallel work of research and industrial projects and their direct connecting points.

References

1. Bullinger, H.-J., ten Hompel, M. (eds.): Internet der Dinge. Springer, Berlin (2007)
2. Günthner, W., ten Hompel, M. (eds.): Internet der Dinge in der Intralogistik. Springer, Heidelberg (2010)
3. Weiser, M.: The Computer for the 21st Century. Sci. Am. 265(3), 94–104 (1991)
4. Feldhorst, S., Fiedler, M., Heinemann, M., ten Hompel, M., Krumm, H.: Event-based 3D-Monitoring of Material Flow Systems in Real-Time. In: Proc. of the 8th IEEE International Conference on Industrial Informatics (INDIN 2010), Osaka (2010)
5. Newman, J., Schall, G., van Barakonyi, I., Schürzinger, A., Schmalstieg, D.: Wide-Area Tracking Tools for Augmented Reality. In: Proceedings of the 4th International Conference on Pervasive Computing, Dublin (2006)
6. Mandeljc, R., Perš, J., Kristan, M., Kovačič, S.: An alternative way to calibrate ubisense real-time location system via multi-camera calibration methods. In: Zajc, B., Trost, A. (eds.) Proceedings of the 19th International Electrotechnical and Computer Science Conference, ERK 2010, pp. 249–252 (2010)
7. IoT-A project: Internet of Things Architecture (2010), `http://www.iot-a.eu`
8. smaRTI project: smart reusable transport items (2010), `http://www.smart-rti.de`
9. DyCoNet project: Dynamische Container Netzwerke (2010), `http://www.dyconet.de/p/about-project.html`
10. Ubisense, `http://www.ubisense.net/en`
11. Mojix, `http://www.mojix.com`
12. Sadr, R.: United States Patent US 7,633,377 B2 (2009)
13. Lufthansa Cargo, `http://lufthansa-cargo.com/de_de`
14. AZM, `http://www.rfid-azm.de`
15. Fraunhofer, `http://www.oid.fraunhofer.de/de/Referenzen.html`

Using RFID in License Plates and Vignettes for Electronic Vehicle Identification

Structured Testing of Passive UHF Systems for Vehicle Identification

Dennis Brandwein[1], Dieter Uckelmann[2], and Björn Beenken[3]

[1] BIBA, Bremer Institut für Produktion und Logistik GmbH at the University of Bremen, Hochschulring 20, 28359 Bremen, Germany
bra@biba.uni-bremen.de
[2] LogDynamics Lab, University of Bremen, Hochschulring 20, 28359 Bremen, Germany
uck@biba.uni-bremen.de
[3] J.H. Tönnjes E.A.S.T. GmbH & Co. KG, Syker Strasse 201, 27751 Delmenhorst, Germany
b.beenken@toennjes.com

Abstract. In Cooperation with the J.H. Tönnjes E.A.S.T. GmbH & Co. KG, the BIBA – Bremer Institut für Produktion und Logistik GmbH conducted a project on electronic vehicle identification via RFID at the University of Bremen. The aim of this research was to find the optimal combination of transponder and reading device for a passive RFID system in UHF frequency band and to suggest improvements. Within the tests, RFID vignettes for windshields and RFID license plates of the Tönnjes Group were analysed and optimised. In a follow-up project the improvements were verified.

Keywords: Electronic Vehicle Identification, Structured Testing, Passive UHF Systems, Windshield Label, RFID License Plate.

1 Introduction

The usage of transponders for electronic vehicle identification constitutes a growing field of application in RFID technology. The usage options range from vehicle access control to vehicle localisation in a delimited area to electronic toll collection or even complex traffic control [1]. For practical reasons, a wide range of these applications has been implemented with active transponders in UHF and microwave frequency ranges so far. Their integrated power supply enables transponders to actively send their own high frequency radio signal, which can be received by corresponding reading devices, even at long distances. By contrast, passive transponders are not battery-powered and, therefore, limited in their reading range. However, passive power supply also entails advantages. Passive transponders can be more compact and are available at significantly lower costs. Moreover, they are more durable, since

Dieter Uckelmann et al. (Eds.): ImViReLL 2012, CCIS 282, pp. 148–155, 2012.

passive transponders are not depending on battery life. To benefit from this in vehicle identification, it is necessary to demonstrate the performance of passive UHF systems under adverse conditions, such as high speeds.

Within the collaborative project EVIdent, the BIBA – Bremer Institut für Produktion und Logistik GmbH and the J.H. Tönnjes E.A.S.T. GmbH & Co. KG tested the applicability of passive RFID systems for electronic vehicle identification. All tested RFID components were operated within the UHF frequency range of 860-960 MHz and supported the air interface protocol according to ISO/IEC 18000-6C respectively EPCglobal Class 1 Gen 2. Because some data is of a confidential nature, not all results of this industrial project can be published.

For an efficient study design, the tests were structured in two phases. In this way, a reasonable number of tests could be conducted. At the beginning, static tests were carried out under defined laboratory conditions in an anechoic chamber. Subject to testing were, on the one hand, RFID vignettes containing special RFID inlays and holograms, which were applied to the windshield. On the other hand, the performance of "transmitting" license plates, internally developed by the Tönnjes Group, was tested. These RFID license plates are equipped with a slot antenna and a passive RFID chip module. Fig. 1 gives an outline of both described types and their attachment to a car. Prior research on RFID license plates has either focussed on active RFID chip modules [2] or standard passive tags were located behind, embedded between or attached in front of the license plates [3].

Fig. 1. Generic Outline of RFID Vignettes and RFID License Plates Attached to a Car

Afterwards, the gained findings were integrated in realistic, dynamic tests. In addition to RFID vignettes and license plates, various RFID reading devices with circular antenna polarisation were compared to ensure a reliable identification at a speed of 120 km/h (due to customer requirements) as well as at a distance of several metres. For each test, five different types of inlays, license plates, handheld readers and reading devices were selected and analysed.

2 Testing Product Quality

Within first test series, the product quality of selected inlays was tested for consistency. Passive inlays are usually mass-produced goods, which are distributed on rolls in large numbers. Therefore, production-related differences in quality cannot be excluded. Earlier studies, such as [4], have shown that some transponders had large variations in their responding qualities. For testing, all available samples of RFID vignettes and RFID license plates were analysed. The arithmetic averages as well as the variance of their responding qualities were calculated. Furthermore, the effect of different samples of auto glass and holograms on the inlays' responding quality was tested. To enable the testing of a large number of transponders with sufficient accuracy, the power output of the reading device was gradually increased from 15 dBm up to 30 dBm.

For this purpose, a test station with a defined distance of one metre between reading antenna and test object was used in the institutes' outdoor area. To broadly eliminate influences by reflections and electromagnetic sources of interference and, consequently, achieve conditions similar to laboratory, the antenna was positioned in a way that enables an upward radiation into the open field. The test station is made of polystyrene plates, which have a neglectable influence on electromagnetic waves.

As a result it should be noted that no significant effect on identification by different types of auto glass could be measured. Sole exception in testing was a metallised windshield; as expected, no identification was possible with an RFID vignette behind, due to the shielding effect of the metal film.

The variance in responding quality of inlays and license plates could be measured: it lies between 0.5 dB and about 2.2 dB. Due to measuring inaccuracies, a tolerance of 1 dB was assumed, whereby ultimately all transponders were admitted to the following tests.

3 Testing Properties in Laboratory Environment

The measurements in the anechoic chamber enabled the analysis of reliable causal relationships. Furthermore, the shielding of the non-reflective room enabled a testing of frequencies that are not permitted in the EU. In this way, the minimal activation power of each transponder type could be measured in dependence of different positions and frequencies.

The activation power of a transponder describes the required energy for operating it and allows drawing conclusions about reading range. The radiated power of the RFID antenna loses intensity with increasing distance to the antenna. According to this, the lower the required power for operating a transponder, the higher is its identification range. Both, test setup and execution, essentially base on the standardised EPCglobal Static Test Method [5], in which two reading antennas are operated bistatically. This method requires a separation of transmitting and receiving antenna to measure the transponders' actual activation power.

In this way, the transmitting power can be reduced by an attenuator without simultaneously damping the receiving performance. Additionally, elements of the ISO/IEC TR 18046 [6] and VDI/AIM 4472-10 [7] standards were taken into account.

The influence of frequency on activation power was nearly identical for all inlays and license plates. Compared to the tested European frequency at 866 MHz, identification at 920 MHz lead to a performance improvement, which was partly distinct with measured values up to -5 dB. Consequently, a fine tuning of the transponder antenna for European frequency range would lead to higher reading ranges. All RFID vignettes reach their minimal activation power at an orientation of their antenna face vertical to the radiation direction of the reading antenna. A rotation by a positive or negative angle leads to an increase in necessary activation power to the same extend. In contrast, RFID license plates have a preferred side due to their asymmetric slot antenna layout, from which they can be identified considerably better. The results of the anechoic chamber resemble those of the previously conducted tests in the outdoor area. This applies to performance variance within one type as well as performance difference between different types. Thereby, the capability of the test setup in the outdoor area could be verified and no further tests in the anechoic chamber (e.g., for windshield inlays that arrived late from manufacturers) were necessary.

As a transition from static to dynamic tests, the performance of different handheld readers was evaluated by their reading range in an open field trial. In this test, no guidance or fixation was set for the handheld readers. Instead, it was moved freely by a person at the level of the transponder. The test is oriented towards applications such as checking parking permissions. The inlays were applied to auto glass and identified by the handheld readers one after another. For identifying the reading range, the distance between handheld reader and transponder was extended until identification was no longer possible. The range was measured with a laser distance meter. In this way, a fast and direct comparison of the handheld readers with sufficient accuracy was enabled.

For handheld readers, enormous differences in performance were measured. However, due to different types and transmitting powers of the installed RFID modules, these findings have to be considered differentially. The two highest-performing tested handheld readers are the largest and heaviest ones as well. They are equipped with a large antenna that requires extended housings. Moreover, the corresponding RFID modules provided the highest power outputs with 27 dBm each.

4 Determination of the Optimal Position

After characteristics and properties of RFID inlays and license plates had been analysed, the optimal position on windshield had to be determined for the inlays. For this purpose, a high-quality measuring device was applied. With the CISC RFID Field Recorder [8] and corresponding sensor modules it is possible to measure the actually received power of a reading device. The minimal activation power measured in the anechoic chamber is used as a reference value of the power required for activation (cf. Fig. 2).

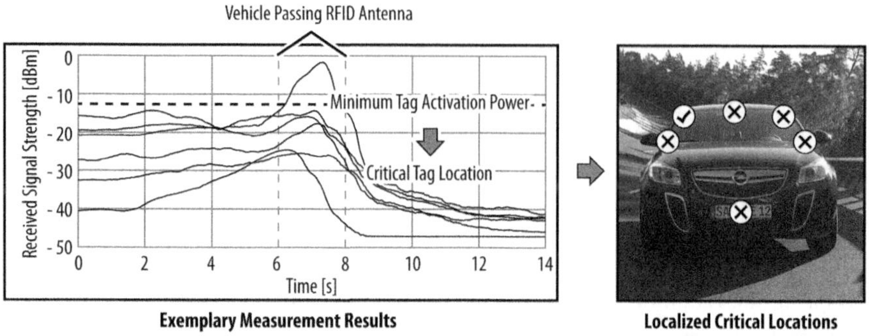

Fig. 2. Generic Comparison of Minimal Activation Power and Measured Receiving Performance

Within this test, different positions at the vehicle were tested to find out where the power was sufficient to operate the respective transponder.

To determinate the optimal position, five sensors were applied to the windshield and another one to the license plate of a test vehicle. This vehicle was driven through an RFID gate with a speed of 20 km/h. The sensor position with the highest measured signal strength was interpreted as the position with the highest probability for identification. The results show that positions in the upper area of the windshield are preferable to lower ones. Accordingly, these positions were chosen for following practical tests.

Furthermore, the effects of different antenna orientations could be derived from the measured values of this test, even though the angles were changing continuously due to the relative motion of the vehicle. If a side-mounted antenna radiates across the road (90°), the vehicle will be in the reading field of the reading device for only a very short time. Optimal reading results were measured at angles between 30° and 60°. The findings are identical for the antenna's tilt angle above the road.

5 Dynamic Tests in Outside Laboratory Environments

Finally, tests were conducted under realistic conditions at the Contidrom Proving Ground and the Ganderkesee Atlas Airfield. Nonetheless, these have to be considered as controlled laboratory environments as tests in real traffic scenarios require increased efforts and security including road blocking, for example, when antenna positions are adjusted.

For electronic toll collection, a reliable identification is essential. The findings of the previous test, such as the optimal antenna position, were considered in the setup for dynamic testing. For performance determination of reading devices, antenna configuration as well as vehicle speed was varied. In function testing and within first test series, single vehicles were identified. Based on the bulk scenarios for electronic toll collection, which are described in the CEN ISO/TS 14907-1 standard [9], typical traffic situations, such as changing lanes and slipstreaming, were simulated afterwards. Three vehicles were driving on two lanes in the same direction with a small distance to each other. To realistically depict the traffic, different vehicle categories were used. Moreover, speed was varied from 10 km/h up to a maximum of 120 km/h on the Airfield and 180 km/h at the Contidrom to meet possible application cases for electronic vehicle identification. An excerpt of the test findings is illustrated in Fig. 3.

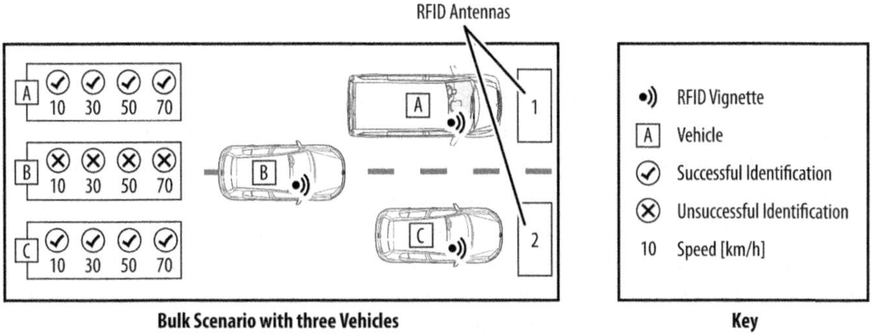

Fig. 3. Exemplary Measurement Results of a Bulk Scenario

Fig. 3 shows that an additional antenna is necessary for capturing a vehicle during lane changes (vehicle B), which is positioned between antenna 1 and 2. Alternatively, antennas with a wider radiation angle can be used. Transponder identification by reading antennas was successful, above as well as lateral to the road. At a height of 3.25 m, all inlays were read at different speeds. At a height of 5.5 m, an identification of the inlays was only possible to a limited extend. For the license plates, identification from above turned out to be more useful than readings from antennas mounted to the left or right. However, for access controls or toll stations, a lateral identification is certainly conceivable, due to separation of the lanes.

6 Results from the First Test Series

The applicability of passive UHF technology for electronic vehicle identification was proved in the tested scenarios. Depending on antenna position and installation height, an identification of inlays and RFID license plates is also possible at high speeds of up to 180 km/h.

Based on the tests, the best fixed and mobile UHF readers as well as inlays for RFID vignettes could be selected. Likewise, conclusions could be drawn on reading performance by varied antenna orientation and positioning. For RFID vignettes, the effect of distance between inlay and hologram could be determined and be taken into account in vignettes' optimisation. Moreover, the gained findings were used in further development of the RFID vehicle identification system of the J.H. Tönnjes E.A.S.T. GmbH & Co. KG, which is already in use in European as well as American markets.

7 Improvements and Second Test Series

The test results and suggestions for improvements were taken seriously by J.H. Tönnjes E.A.S.T. GmbH & Co. KG and lead to further research and optimisation. The development aimed to achieve reliable identification at higher speeds and distances. In October 2011, a new test series to verify the improvements was started. Again, first tests were carried out in the anechoic chamber. The results were promising. Especially the license plate activation levels showed great improvements up to 6 dB. When testing in the outside area, a new problem had to be solved. The testing distance of one metre was not enough to distinguish quality levels of the license plates, as all samples could be read at the minimum level of 15 dBm, even though different antenna layout options and slot casting were used. Consequently, the reading distance was increased to two metres to analyse which sample worked best.

The most exciting tests, however, were the high speed tests at 260 km/h and more. For these tests, the Opel Test Center in Dudenhofen offered their support. A special Opel Insignia OPC Unlimited was equipped with an RFID vignette and license plate. The RFID vignette had to be placed in the non-metallised area of the windshield close to the mirror to provide readability. However, due to the shielding, read rate problems could be expected. For reference, additional transponders were attached to the car. The positioning of the RFID antennas at the bridge spanning across the high speed test ground required special attention. As the road is tilted at a degree of 35°, the three reading antennas had to be fixed with the same tilt level to ensure consistent reading distances (Fig 4).

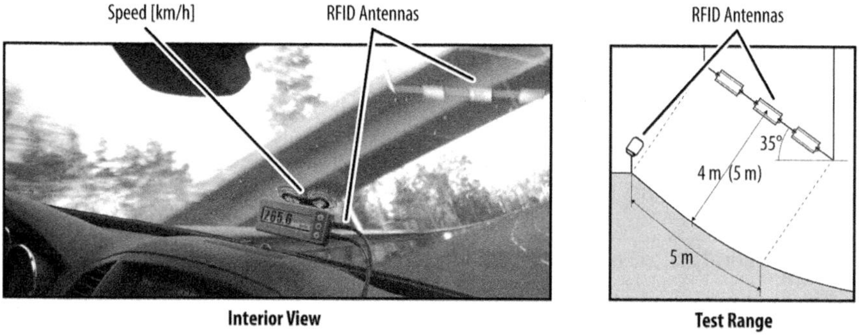

Fig. 4. High Speed Test Range

Additionally, one antenna was attached to the left of the high speed lane. All four antennas were switched automatically. Tests were carried out at speeds of 80 km/h to 260 km/h in steps of 20 km/h with three rounds at each speed. In this case, the RFID license plates achieved 100% reading rates at all speeds. Even the RFID vignette in the non-metallised area of the windshield achieved 69% more reads than the reference transponders. Due to the small number of test runs at each speed in an outdoor lab-environment, it would be good to verify the results in the first phase of a real implementation.

At the time of writing, the repeated bulk scenario tests at Atlas Airfield still need to be carried out. However, the test results so far show considerable improvements and a reliable reading of the license plates at all speeds. Therefore, and due to the expected speeds of the possible applications of electronic vehicle identification, license plates are not only suitable for access control and toll collection, but also for traffic control concepts. Although the results from both test series are promising, the usage of passive tags on metallised windshields is and remains a challenge for further research.

References

1. Blythe, P.: RFID for road tolling, road-use pricing and vehicle access control. In: IEE Colloquium on RFID Technology 1999/123, pp. 8/1–8/16 (1999)
2. Zhao, W., Wang, G., Lai, X.: Active E-Plate with Slot Antenna. In: 4th International Conference on Wireless Communications, Networking and Mobile Computing 2008, pp. 1–3 (2008)
3. Mosby Jr., D.L.: Embedded Radio Frequency Identification Device License Plates for Roadside Use in Nebraska. Dissertation, University of Nebraska - Lincoln (2010)
4. Ramakrishnan, K.M., Deavours, D.D.: Performance benchmarks for passive UHF RFID tags. In: German, R., Heindl, A. (eds.) Proceedings 13th GI/ITG Conference on Measurement, Modeling, and Evaluation of Computer and Communication Systems, Berlin, Offenbach, pp. 137–154 (2006)
5. EPCglobal: Static Test Method for Applied Tag Performance Testing, `http://indicod-ecr.it/documenti/epc/static-test-method.pdf`
6. ISO: ISO/IEC TR 18046 - Information technology - Automatic identification and data capture techniques - Radio frequency identification device performance test methods. Geneva (2005)
7. Verein Deutscher Ingenieure e.V.: VDI/AIM 4472 Blatt 10 - Anforderungen an Transpondersysteme zum Einsatz in der Supply Chain - Testverfahren zur Überprüfung der Leistungsfähigkeit von Transpondersystemen (RFID). Düsseldorf (2008)
8. CISC Semiconductor Design+Consulting GmbH, `http://www.cisc.at/field-recorder-new.html`
9. ISO: CEN ISO/TS 14907-1 - Road transport and traffic telematics - Electronic fee collection - Test procedures for user and fixed equipment - Part 1: Description of test procedures (2010)

The Internet of Drinks

Lab-Based Research for the Taste of It

Marco Lewandowski[1], Kolja Schmidt[2], Christoph Kielhorn[2], and Dieter Uckelmann[2]

[1] BIBA - Bremer Institut für Produktion und Logistik GmbH at the University of Bremen, Hochschulring 20, 28359 Bremen, Germany
lew@biba.uni-bremen.de
[2] University of Bremen, c/o BIBA GmbH, Hochschulring 20, 28359 Bremen, Germany
{sdt,uck}@biba.uni-bremen.de

Abstract. It is not always easy to depict the benefits of RFID and the Internet of Things. Especially in research, it is sometimes challenging to find illustrative scenarios that grab the attention of the audience and convey a positive message of technology usage. Fortunately, researchers around the world are imaginative enough to come up with creative and sometimes – at least in the first impression – funny approaches. This humorous effect should not be underestimated in a world where new technologies raise a lot of fear, uncertainty and doubt. The "Internet of Drinks" is an obvious play on words related to the Internet of Things. This paper will describe some example scenarios that utilize Internet of Things technologies to produce, control and monitor beverages and will show some obvious and easy to understand demonstrators.

Keywords: Internet of Things, Automation, Sensor Networks, RFID.

1 Introduction

A classic example of utilizing a beverage scenario to illustrate a new technology is the *Trojan Room Coffee Pot* at the University of Cambridge [1], which became famous in 1991 as a first application to utilize a webcam. With the development of ubiquitous computing, the "Smart Coffee Machine" seems to be a very motivating topic for researchers. It includes automation effects, such as activating the coffee-machine when leaving the bed after 7 a.m. [3]. Coffee capsules may communicate with the coffee machine via barcodes [4]. We even see Internet accessible coffee makers, which even may be misused as a security leak into connected Windows systems [5]. Consequently, there is a book entitled *"The Spy in the Coffee Machine: The End of Privacy as We Know It"* [6]. Furthermore, other beverages have inspired researchers and practitioners alike. RFID is being used to track beer kegs [7]. Coca Cola supplies RFID-based dispensers to provide a higher level of customer choice and to collect consumption data over a private wireless network [8]. Self-service "pour-your-drink" stations have been developed [9]. But why should people order a refill, if automatic liquid level detection can automate this process [10]? In case of alcoholic beverages,

Dieter Uckelmann et al. (Eds.): ImViReLL 2012, CCIS 282, pp. 156–166, 2012.

RFID-wristbands may be used to verify a customer's age [11]. Automated wireless beverage tracking solutions [see, e.g., 12, 13, 14], ensure that beverage shrinkage is avoided. The workflow of producing a cocktail has been analysed [2]. Bartenders gestures have even been used for mobile interaction systems [15].

A question to ask is: Why are beverages so interesting to researchers in ubiquitous computing, practitioners and end-users alike? For the coffee maker, one answer lies in the so called "Smart Home" that includes connectivity and control of everyday objects (see, e.g., [16]). A second reason is that a coffee machine is one of the best demonstrators for ubiquitous computing, as it does not require explanation itself and it has a positive connotation. The latter is also true for cocktails.

In this paper the integration of Internet-based monitoring and control in these two different beverage scenarios will be shown. Demonstrator one will show a fill level sensing application that uses Pachube[1] – an Internet application that allows flexible integration of sensors for defined locations – to monitor a milk carton at a cappuccino machine, whereas in demonstrator two a self-service cocktail machine, that allows access through RFID-cards and individual mixtures through a web-based interface is shown.

2 Demo-Case 1: Monitoring the Fill Level of a Milk Carton for a Cappuccino Machine

The following section shows how easily sensors can be integrated into the Internet of Things in a prototypic application. As an example, the fill level of a customary milk carton is measured, for hygienic reasons, by its weight m (in kg) and made retrievable via Internet. A technical development is described to process the physical information of the carton's weight as well as wirelessly transfer them into the Internet and make them retrievable there.

2.1 Developing a Sensor Integration

There are different methods for sensory acquisition of a fill level. Technically, the measurement can be done mechanical (e.g., by measuring the differential pressure) and/or electrical (e.g., by measuring the capacity between two electrodes). In the following, a hybrid approach of both mentioned methods will be shown.

A resistance change, resulting from a change in weight due to a fill level change, can be measured sufficiently accurate by a force-sensitive resistance (FSR). In detail, an FSR varies its electrical resistance depending on the pressure that is applied on the active sensor area.

In this example application, a square FSR of Sparkfun Electronics with an active sensor area of 437 x 437 mm was chosen. The advantage of a square FSR is that even in edge areas a reliable resistance change can be detected. Thereby, positional changes of the beverage carton can be compensated by the sensor.

---

[1] `www.pachube.com`

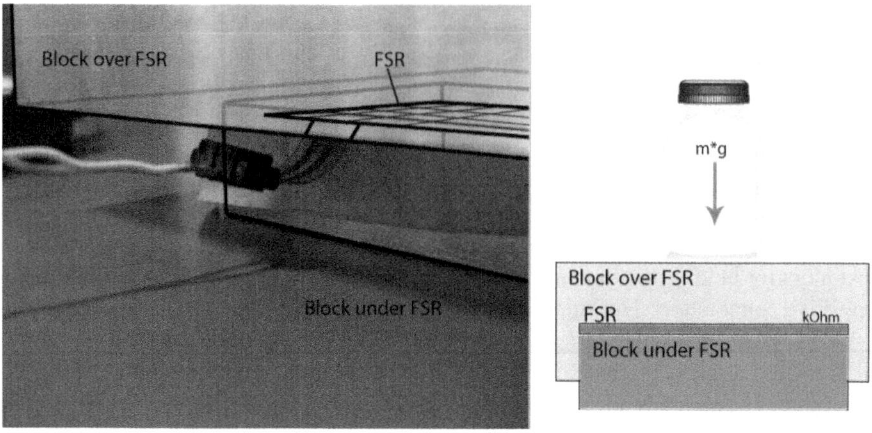

Fig. 1. Integration of the FSR

The active sensor area of the FSR is floating between two aluminium blocks. The beverage carton to be measured is placed on the upper block (see Fig. 1). Its weight force is fed into the upper block, which affects the sensor area. Since the lower block is not floating, the FSR is pressed between the two aluminium blocks, whereby a resistance change is generated.

For acquisitioning the resistance changes of the FSR and communicating with the Internet, a microcontroller is required. The so-called Arduino[2] platform constitutes an economic, flexible and sufficiently powerful system. It consists of hard- and software components and is open source.

The core of an Arduino is an ATMega328 microcontroller, an input/output board with analogue and digital in- and outputs. These can be connected to, for example, sensors or actuators. The development environment is based on Wiring, an easy-to-learn Java dialect. There is the possibility to extend the functionality of an Arduino (e.g., for communicating with the Internet) by extending the system with so-called shields. For the chosen application site, only the wireless local area network (WLAN) was applicable, so a WIFLY shield was used. The WIFLY shield bases on an RN-131G WLAN module of Roving Networks, an SC16IS750 SPI-to-UART controller and a dot matrix prototype for internal circuits. It can be attached to an Arduino and connects it to an 802.11b/g WLAN network with WEP and WPA2 encryption. The module uses the Vin Pin of the Arduino for power supply and is equipped with a 3.3 V voltage regulator. It is actuated by the serial communication interface of the Arduino or, more precisely, its corresponding pins. For operating, a 3.3 V power supply and three digital inputs are required. Consequently, only three of the 24 in- or outputs are internally assigned. The WIFLY shield is connected to the local WLAN network and communicates via HTTP protocol. The corresponding library for programming is open source and is stored in the Arduino.

[2] `http://www.arduino.cc/`

The power supply of the Arduino itself is provided by a 9 V power adapter through a customary 3.5 mm jack input. The programming of the Arduino is done by use of the mentioned software, which programmes the microcontroller by a so-called in-system programmer (ISP). As a protection against splashing water and dust, the system is encapsulated in a box. The connecting cables for the sensor run out of it and are isolated.

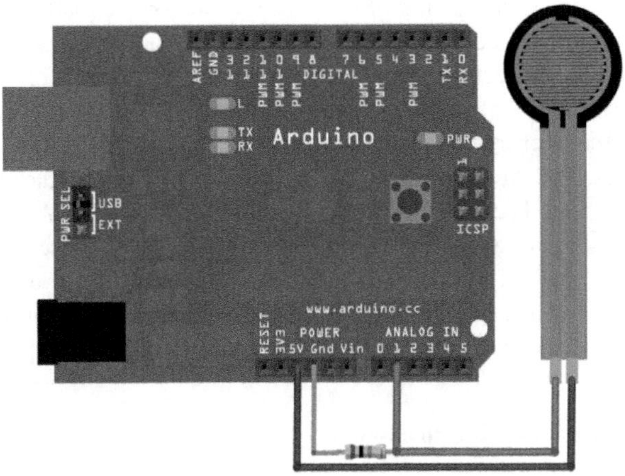

Fig. 2. Circuit diagram with Arduino and FSR

Fig. 2 illustrates the implemented circuit. It shows the Arduino connected to the chosen FSR (right in the figure). The positive pole of the FSR is connected to the 5 V power supply of the Arduino (WIFLY). The negative pole of the FSR is connected parallel: on the one hand, against mass, to a 10 kilo-ohm carbon film resistor and, on the other hand, to the analogue input A1. The measurement takes place through a voltage divider. This means that the voltage drop of the FSR is measured in relation to a 10 kilo-ohm resistance. The corresponding formula is:

$$V\_0=V\_cc*(R/(R + FSR))=5V*((10000\,\Omega)/(10000\,\Omega + FSR\,\Omega)) \quad [V; Volt]. \tag{1}$$

The explained method for resistance measurement was chosen, because the analogue input of the Arduino is only able to measure potentials and convert them digitally (by a 10 Bit analogue/digital converter). A voltage of 5 V corresponds to a digital value of 1023; correspondingly, a voltage of 0 V corresponds to a digital value of 0. The system's measurement resolution is not high, due to the lacking of decimal places, but it is adequate for the desired purpose. Based on the digital voltage value, the fill level is calculated. A maximally filled beverage carton has a weight of 1.05 to 1.06 kg. This corresponds to a digital value of 510. An entirely empty beverage carton has a weight of ca. 0.03 kg. This corresponds to a digital value of ca. 180 for the Arduino (including the weight of the aluminium block). Consequently, 180 to 510 are the approximate limits for measurement.

2.2 Integration into the Internet

The Arduino processes the digital measurement value and prepares it for being transmitted by the WLAN network. Therefore, the already mentioned free library is used for the WIFLY shield, which allows the Arduino to act as a client that is able to communicate with the Internet via HTTP protocol. Through POST commands, the corresponding API key, date and sensor measurement value are transferred.

The measurement values of the FSR are sent to the online platform "pachube.com". Pachube is an online service, which enables storing and reading sensor data. It is based on the idea not only to connect servers in the Internet, but also objects or things. The online service offers a technical infrastructure and an alternative for an Internet of Things.

Pachube is free of charge up to a certain volume of use and functions through in- and output interfaces. The Arduino sends the measurement values by the WIFLY shield as an output through the Pachube API to the input interface of the online platform (in detail, in a so-called input feed). The sending interval amounts to 30 seconds, since shorter cycles would be charged by the online platform. The transmitted data can be viewed graphically at a so-called feed page provided by Pachube. For further application cases, the sensor data can also be integrated in other systems through an output feed.

2.3 Calibration and Measurements

For testing the overall system, 1 litre water bottles with varying fill levels were used to simulate different load levels. For this purpose, 3 water bottles were placed on the sensor area at an interval of ca. 30 minutes. The first bottle had a weight of 1,036 kg and was entirely filled. It generated a quite constant digital sensor value of ca. 501 (see Fig. 3). The second bottle was a little more than half-full and had a weight of 0,595 kg. The digital value of 360 was stable over time as well. Finally, a bottle with a slightly lower fill level was tested to reveal the differences in consumption of small units. It had a weight of 0,512 kg and reached a digital sensor value of ca. 305. In neutral, unloaded state, the system shows a value of 180, due to the already mentioned upper laid aluminium block. The respective digital values over time are illustrated in Fig. 3.

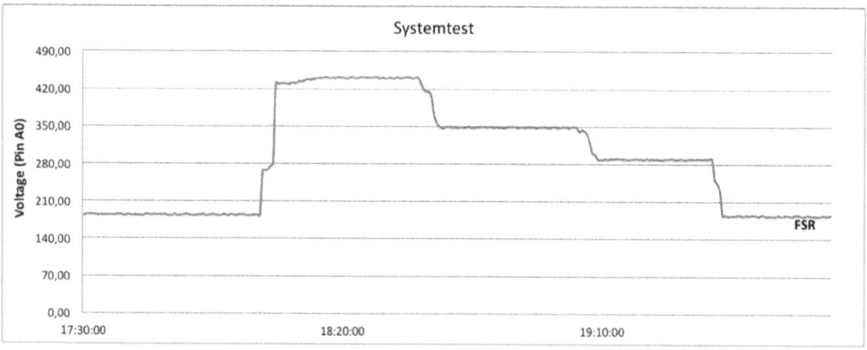

Fig. 3. Testing the overall system with varying weights

In the context of connecting things to the Internet, this scenario showed how physical information (fill level) can be integrated into the internet. A prototype was developed (see Fig. 4), which is able to transfer information through a local WLAN by HTTP protocol to the online platform Pachube. There, the transferred sensor values are graphically presented and can be integrated in diverse additional applications.

Fig. 4. Final illustration of sensor integration

3 Demo-Case 2: The Cocktail Factory

The Cocktail Factory is a second demonstrator, utilizing a combination of certain state-of-the-art technologies to demonstrate aspects of the Internet of Things idea. The system produces cocktails made of basic ingredients based on a recipe, which is saved on local RFID transponders outside the machine. Automation technology and controlling software, beverage processing technologies together with RFID and 2-dimensional barcode have been engineered to provide an automated Cocktail Factory that is currently in the transition stage from lab-based research to a valid business offering.

3.1 Structure of the System

The system consists of several components, which all belong to the categories housing, beverage production, or device control. A general overview is provided in Fig. 5.

One of the key components of the beverage processing unit includes the tanks (4), which hold the basic ingredients. Each tank is connected by hoses to a food-grade diaphragm pump. The diaphragm pumps are permanently supplied with electrical power and pump against a closed valve. An integrated pressure switch turns off the pump when the required pressure between the pump and valve is established. Hoses, pumps and valves build the beverage processing unit (6) and end up in the dispensing unit (1). The beverage processing unit is engineered according to the standards of European Standard DIN EN 1672-2 [17] and the German standard for dispensing equipment DIN 6650 [18].

Essentially for the demonstration of technologies related to the Internet of Things, the machine is operated via decentralised data storage solutions like RFID, 2-dimensional barcode and, optionally, via a web-based interface to adjust the recipes. There is no way and no need to control the machine via an advanced graphical user interface, for instance applying a touch screen. Accordingly, the key elements of the storage unit are the RFID transponders (2), the 2-dimensional barcodes and the reader unit (3) as well as a Programmable Logic Controller (PLC) inside the corpus (5) of the machine.

The simple principle of mixing drinks with the machine is the time-controlled opening of the valves. The liquid transport is automatically achieved by decentralized control loops with integrated pressure switches in the pumps. The low cycle time of the PLC allows very precise sampling volumes from the tanks while maintaining a high speed of the mixing process. The respective timetables for the valves are stored as recipes on the data carriers.

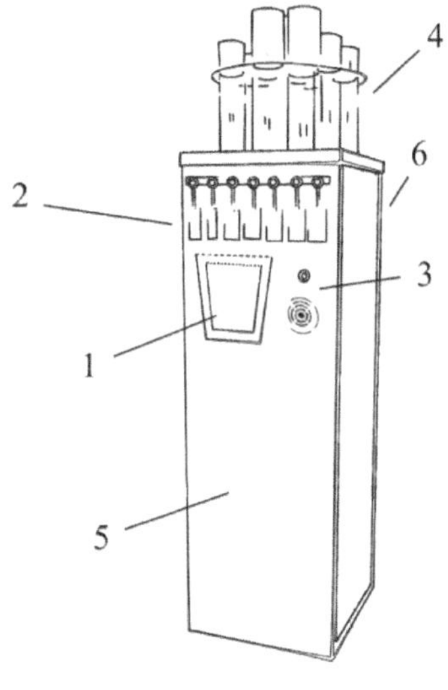

1 dispensing unit
2 RFID transponder
3 reader unit
4 storage tank
5 corpus
6 beverage processing unit

Fig. 5. General structure of the system

Regarding the components in detail, the RFID reading unit is a HF-module that operates with 13.56 MHz standard transponders. It is connected to the PLC via RS232. The PLC is a Windows CE-based embedded computer running a real-time PLC and is programmed according to the standards of DIN IEC 61131-3 [19]. All digital inputs and outputs as well as the RS232 connection are handled via an EtherCAT fieldbus system, which is well established in automation technology. A second reading unit is used to enable the reading of recipes, which are saved based on 2-dimensional barcodes (e.g., QR-codes). It is also connected to the fieldbus via RS232. To prevent abuse of control or reading failures by incorrect codes on the data carriers, recipes are encoded and are provided with a check sum.

3.2 Operation

The use of Auto-ID technologies permits a hitherto unusual operation of the machine. The RFID transponders act as a storage device for individual recipes and as a trigger for the ordering process.

The potential customer reads the cocktail description on the RFID transponder – a smart card that has enough space for human readable information. He decides for a specific cocktail based on the written information that includes the ingredients and the name of a drink. Instead of remembering this drink now and ordering, for example based on a typical touch-screen, the physical RFID card serves as the information carrier for the order. When it is moved in the reading field of the machine, it starts the production process. Fig. 6 below gives an overview of the process.

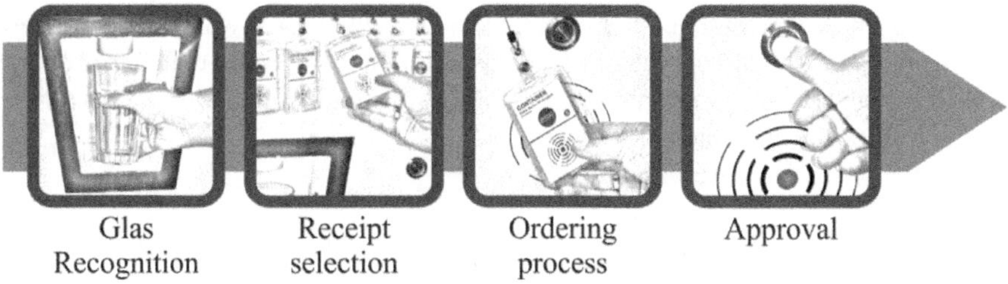

Fig. 6. Process of ordering via RFID

Another scenario is to operate the machine via a smart phone app. This allows selecting a cocktail, the adjustment to the personal taste, and the creation of a completely own beverage with a smart phone software. In this case, the phone itself has to serve as a carrier of information regarding the reader field of the machine and the according production triggering. This has been solved using a dynamically generated 2-dimensional barcode with QR technology directly on the phone's screen. This code contains the recipe so that the machine is programmed accordingly and triggers the production. Fig. 7 illustrates the ordering process via smart phone app. A similar scenario would be possible based on Near Field Communication (NFC) in the future. Electronic payment solutions through mobile phones have not yet been looked at, but could easily be integrated.

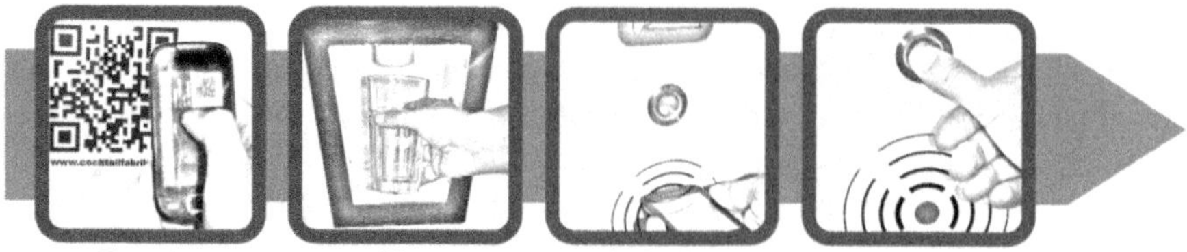

Receipt selection and adjustment via mobile phone | Glas recognition | Ordering process via QR code on the mobile phone | Approval

Fig. 7. Process of ordering via smart phone app

In both cases, the interaction with the machine gives the user a haptic feeling regarding the ordering process. Necessary interactions with the machine are minimized. A feedback of the command that is given via the data carriers is based on light signals.

3.3 Impact on the Business Usage and Marketing Strategy

Mixed drinks are enjoying the hospitality of an ever-increasing popularity. For business owners, cocktails are an interesting product with high profit margins, provided that a high quality level concerning the cocktails is given.

The control is characterized by an innovative user interface, which allows meeting the conditions of a minimalist user input, based on the configured chip cards or smart phone app. This opens up interesting possibilities of marketing and usage in different areas of event catering, hotels and gyms.

Essential characteristics of the device are:

- Automated and accurate mixing of basic ingredients
- Control via electronic chip card (RFID) and QR-code
- Dynamic adjustments of recipes with different ingredients.

The storage of recipes directly on the RFID card includes the ability to use RFID cards as "cocktail vouchers". There are possible applications to create personal drinks so that cocktail competitions could be organized. For fairs and exhibitions the Cocktail Factory allows stimulating high visitor frequencies to exhibitor booths.

3.4 Issues Regarding the Internet of Things and Laboratory-Based Research and Teaching

The example shows an interesting laboratory-based development to demonstrate various state-of-the-art techniques. The system is particularly illustrative and, therefore, suitable to demonstrate the relationships between distributed control and data storage with the actual control of industrial control technology. This example motivated researchers and students to work overtime to meet deadlines, which, as a first proof of concept, happened to be a Christmas party.

As described in the introduction, the simple but realistic scenario of an automated beverage production is of great interest in application-oriented research. In particular, the general public easily understands research ideas such as decentralized control and the Internet of Things using this demonstrator. This also applies to the used technologies themselves, which are still not ubiquitous in everyday life. In the context of higher education, the demonstrator also allows an in-depth introduction to the methodology and technology for students.

4 Conclusion and Outlook

The two beverage scenarios prove that in lab-based research, the term Internet of Drinks can be used to inspire the usage of ubiquitous and automation technologies among students and researchers. It allows explaining difficult research concepts and technologies in well-known scenarios to end-users and it triggers a broad end-user audience on events. Different senses – seeing, touching, tasting – together with a sense of humour are stimulated, thus leading to a higher human cognition and remembering. Eventually, it may foster technology adoption in other fields of applications.

References

1. Fraser, Q.: The Trojan Room Coffee Pot: A (non-technical) biography, `http://www.cl.cam.ac.uk/coffee/qsf/coffee.html`
2. William Brace, T.: Life Unwired The Future of Telecommunication and Networks. In: Neuvo, Y., Ylönen, S. (eds.) Bit Bang Rays to the Future, pp. 42–62. Helsinki University Press, Helsinki (2009)
3. Tassimo: TASSIMO Home Brewing Systems, `http://www.tassimo.ca/tassimo/page?siteid=tassimo-prd&locale=caen1&PagecRef=170`
4. Churchill, E.F.: Ps and Qs: On Trusting Your Socks to Find Each Other. Interact. 2(16), 26–32 (2009)
5. O'Hara, K., Shadbolt, N.: The Spy in the Coffee Machine: The End of Privacy as We Know It. Oneworld Publications, Oxford (2008)
6. Lynch, K.: RFID and the Beverage Marketplace (part 1), `http://rfid.thingmagic.com/rfid-blog/bid/39173/RFID-and-the-Beverage-Marketplace-part-1`
7. Hayes Weier, M.: Coke's RFID-Based Dispensers Redefine Business Intelligence, `http://www.informationweek.com/news/mobility/RFID/217701971`
8. Swedberg, C.: Wall of Beer Lets Patrons Draw Their Drinks, `http://www.rfidjournal.com/article/view/8651`
9. Bhattacharyya, R., Florkemeier, C., Sarma, S.: RFID Tag Antenna Based Sensing: Does your Beverage Glass need a Refill? In: 2010 IEEE International Conference on RFID, Orlando, pp. 126–133. IEEE Press, New York (2010)
10. Swedberg, C.: Putting Drinks on the Cuff, `http://www.rfidjournal.com/article/view/987`
11. Capton Inc.: Beverage Tracker RFID Spout, `http://www.rbscrp.com/capton/Beverage%20Tracker%20Overview%20and%20Specs.pdf`

12. Mitsubishi Electric Research Laboratories: Wireless Liquid Level Sensing for Restaurant Applications, http://www.merl.com/papers/docs/TR2002-21.pdf
13. Rock, J.: Beverage Control-Systems: A new look at an old problem. Hosp. Upgrad. Fall, 42–44 (2004)
14. Schreiber, D., Aitenbichler, E., Ständer, M., Hartmann, M., Ali, S.Z., Mülhäuser, M.: The Mundo Method (An Enhanced Bottom-Up Approach for Engineering Ubiquitous Computing Systems. Cepis Upgrad. 12(1), 25–33 (2011)
15. Jong-Woon Yoo, W., Kim, C., Choi, W., Park, K.: Cocktail: Exploiting Bartenders' Gestures for Mobile Interactions. Int. J. Mob. Hum. Comput. Interact. 2(3), 44–57 (2010)
16. Surie, D., Laguionie, O., Pederson, T.: Wireless Sensor Networking of Everyday Objects in a Smart Home Environment. In: International Conference on Intelligent Sensors, Sensor Networks and Information Processing (ISSNIP), pp. 189–194. IEEE Press, New York (2008)
17. DIN EN 1672-2: Food processing machinery, Basic concepts Part 2: Hygiene requirements; German version EN 1672-2:2005+A1 (2009)
18. DIN 6650: Dispense systems for draught beverages (2006)
19. DIN IEC 61131-3: Programmable Controllers – Part 3: Programming languages (2009)

Author Index

GPSR Compliance

The European Union's (EU) General Product Safety Regulation (GPSR) is a set of rules that requires consumer products to be safe and our obligations to ensure this.

If you have any concerns about our products, you can contact us on ProductSafety@springernature.com

In case Publisher is established outside the EU, the EU authorized representative is:

Springer Nature Customer Service Center GmbH
Europaplatz 3
69115 Heidelberg, Germany

Batch number: 10844621

Printed by Printforce, the Netherlands